I0817305

BORMAN LOVELL ANDERS

THE APOLLO PHOTO ARCHIVE

APOLLO 8 IN PHOTOGRAPHS

BORMAN LOVELL ANDERS

THE APOLLO PHOTO ARCHIVE

APOLLO 8 IN PHOTOGRAPHS

BORMAN LOVELL ANDERS

J.L. PICKERING AND JOHN BISNEY

WITH ED HENGEVELD

4880 Lower Valley Road • Atglen, PA 19310

Other Schiffer books by the authors

Apollo 1 in Photographs: The Apollo Photo Archive, 978-0-7643-7009-0

Apollo 7 in Photographs: The Apollo Photo Archive, 978-0-7643-7010-6

Library of Congress Control Number: 2025939997

Designed by Beth Oberholtzer
Cover design by Jack Chappell
Type set in Eurostile/Collier/Arno Pro

ISBN: 978-0-7643-7084-7
ePub: 978-1-5073-0636-9

Printed in India
10 9 8 7 6 5 4 3 2 1

Published by Schiffer Publishing, Ltd.
4880 Lower Valley Road
Atglen, PA 19310
Phone: (610) 593-1777; Fax: (610) 593-2002
Email: info@schifferbooks.com
Web: www.schifferbooks.com

Contents

Foreword

I had just turned six when my dad flew on Apollo 8. My firsthand memories of that time come in bright bits and pieces but completely lack the context of a more sentient, mature perspective. Probably the most interesting "confession" is that when my parents told me in early 1969 that we were moving to the Washington, DC, area, I thought that we would meet all the astronauts who lived in that city, because I had the myopic expectation that all cities had space programs. Having my new classmates so excited that my dad was an astronaut baffled me at first but helped with my appreciation of Apollo's significance.

My understanding of Apollo came mostly from the stories Dad told us, the literature, and the periodicals we collected. In later years, I gained my broadest knowledge of the program from Andrew Chaikin's *A Man on the Moon*, and more recently, *Rocket Men* by Robert Kurson.

I believe that a common human attribute is appreciation of images over the written word, and I am a visual person; so, while written accounts can create imagined pictures in my mind, a significant portion of my current perspective came from the Tom Hanks TV series *From the Earth to the Moon*, which was based on *A Man on the Moon*. Now, this set of books by J. L. Pickering and John Bisney chronicling Apollo in photographs reveals new details of these historic missions and is another excellent contribution to the cultural understanding of them.

One hundred years from now, I predict two things from the Apollo program will be remembered: Neil Armstrong's first step on the Moon, and the unexpected image captured by my dad on Apollo 8, *Earthrise*.

Today I look at the Moon and marvel that I was raised by a human who had been there. I look forward to when we return and effectively establish a continuing presence on that celestial body so close to our own. The twenty-four-hour-a-day view of Earth will be amazing and will help us be better stewards of it. These days, I believe that the most significant scientific discovery of Apollo was the Earth.

While looking through this compendium of images from Apollo 8, I believe your mind will enjoy feasting on many rare or unseen images. But remember, as Dad said, "We went all that way to discover the Moon, and what we really discovered was the Earth."

Lt. Col. Greg Anders, USAFR (Ret.)
Bellingham, Washington
President, Heritage Flight Museum

Anders family in 1968. Clockwise from top left: Bill, Valerie, Greg, Alan, Glen, Gayle, and Eric

Introduction

Apollo 8, the first mission to take astronauts to the Moon and back, established two landmark exploration firsts: the first manned flight to escape Earth's gravity, and the first to orbit another celestial body. Reaching 24,593 mph, the mission also set both manned speed and distance-from-Earth records and went into the history books as one of the more significant overall space achievements.

Apollo 8 had originally been planned as the first flight of the combined Apollo command and service modules (CSM) with a lunar module (LM) in Earth orbit. When the LM manufacturer, Grumman Aircraft Engineering Corp., was unable to deliver an operational spacecraft before the end of 1968, however, NASA boldly changed the flight to be a CSM-only mission to orbit the Moon. The flight would also test the systems, trajectories, and operations for getting there and back, critical for a manned landing. Rumors had circulated that the Soviet Union was preparing to attempt a manned orbit of the Moon as well, adding to the urgency to get there first.

USAF colonel Frank Borman, forty, was commander, while Navy captain James A. Lovell Jr., forty, served as CM pilot, and USAF major William A. Anders, thirty-five, was the rookie LM pilot (though there was no LM). Borman and Lovell had set a manned-spaceflight endurance record together on Gemini VII; Apollo 8 was the first time NASA flew two crewmates together again.

Apollo 8 was the first manned mission launched by a Saturn V, and the first from NASA's LC-39 on Merritt Island at the Kennedy Space Center (KSC), Florida. The flight launched at 7:51 a.m. EST on December 21, 1968, into a low Earth parking orbit to check out its systems. Two hours and fifty minutes later, during the second orbit, the S-IVB stage was reignited for translunar injection.

Apollo 8 then separated from the S-IVB, and the crew rotated the CSM to photograph the spent stage. They then practiced flying in formation. The stage was sent into a solar orbit as Apollo 8 headed toward the Moon.

As they sped away, the crewmen could view and photograph all of Earth for the first time. Huge receivers near Madrid, Spain, and Goldstone, California, allowed live television transmissions from the CM, using a 4.5-pound black-and-white camera identical to one first used on Apollo 7.

Two television transmissions were made en route to the Moon: two were made in lunar orbit; and two were made during the return flight. The first transmission took place on December 22, approximately halfway between Earth and the Moon. The second transmission, on December 23, showed Earth from a distance of 180,000 miles, and the crew sent back the first television pictures of Earth in its entirety. These were relayed to Europe and the Pacific by using NASA's ATS-1 and ATS-3 satellites.

Apollo 8's lunar trajectory required only minor midcourse correction, and on December 24 the spacecraft passed about 70 miles in front of the Moon at a speed of 5,720 mph. On the far side of the Moon, the astronauts fired the SM's Service Propulsion System (SPS) engine at 3:59 a.m. CST to brake the spacecraft into lunar orbit. After flying two elliptical orbits of 193 by 70 miles, the spacecraft was placed in a nearly circular orbit of about 69 miles for eight more orbits—for a total of twenty hours.

When Apollo 8 began its fourth pass across the near side of the Moon, the crew saw an Earthrise for the first time in human history (a sight rarely visible from the lunar surface due to the synchronous rotation of the Moon around Earth).

Anders saw Earth appearing from behind the lunar limb and called to Borman and Lovell, taking a black-and-white photo as he did. Anders then snapped two color images, including the photo that became known as "Earthrise," widely regarded as one of the most historic photos ever taken.

The first of two TV transmissions on Christmas Eve gave audiences their first close-up views of the Moon. The second transmission ended with the three astronauts in turn reciting verses 1 through 10 of chapter 1 of the book of Genesis, from the King James Version of the Bible. At the time, the broadcast was the most watched program in television history, with an audience estimated at one billion people. Intelsat III F-2 relayed the live TV to Europe and South America. Recorded broadcasts reached thirty more countries within twenty-four hours. Borman concluded the transmission with "And from the crew of Apollo 8, we close with good night, good luck, a Merry Christmas, and God bless all of you, all of you on the good Earth."

Just two and a half hours later, at 12:09 a.m. CST on December 25, on the far side of the Moon, Apollo 8's SPS engine was ignited for trans-Earth injection (TEI), to accelerate out of lunar orbit. It was critical—an engine failure would have stranded the men with no hope of rescue. "Please be informed . . . there is a Santa Claus," radioed Lovell when the spacecraft regained contact with Earth. Only one midcourse correction was required during the return flight.

At about 10,357 miles from Earth, the CM separated from the SM. Fifteen minutes later, Apollo 8 reentered Earth's atmosphere at 24,243 mph. The spacecraft splashed down on December 27 in the Pacific Ocean, 1,100 miles south of Hawaii and 3 miles from the aircraft carrier USS *Yorktown*. It was the first splashdown to occur in darkness (at 4:51 a.m. local time), and the crew was not recovered until dawn, forty-three minutes later.

A group of ten Soviet cosmonauts sent a telegram of good wishes to the astronauts, and Soviet chairman Nikolai Podgorny sent a congratulatory cable to President Lyndon Johnson. Johnson called the astronauts to say that he remembered "when we saw Sputnik racing through the skies, and we realized that America had a big job ahead of it. It gave me so much pleasure to know that you men have done a large part of that job."

Although by 1968 NASA's budget was $3.9 billion, down from 1966's high point of $5.93 billion, and Apollo was quickly nearing its zenith, Apollo 8 was a landmark achievement in the space race and in human exploration.

Acknowledgments

Mike Acs
Shane Bell
Helene Brown-Milbert
Jeff Carr
Andrew Chaikin
Sarah Cunningham
Mike Gentry
William Killen
Eugene Kranz
Cindy Lee
Jennifer Levasseur
Sy Liebergot
Jim Ragusa
Bob Sieck
Stephen Slater
Myra Taub Specht
Ron Specht
Liza Talbot
Roy Tharpe
Mark Usciak
Tom Usciak
Ron Woods

CHAPTER 1

The Astronauts

Unpublished 1964 portrait of Maj. Frank F. Borman II, USAF, at the Manned Spacecraft Center (MSC) in Houston, Texas, by photographer John Holland

The Apollo 8 emblem was designed by command module (CM) pilot Jim Lovell, with final artwork by National Aeronautics and Space Administration (NASA) artist William Bradley. Lovell sketched out his design while returning to Houston from training in California after learning the mission had changed from an Earth-orbital to a circumlunar flight.

Unpublished 1964 MSC portrait of Lt. Cmdr. James A. Lovell Jr., USN, who wears his US Naval Academy ring. Holland often inserts white paper shirt cuffs into the subject's jacket sleeves for his portraits.

Unpublished 1964 MSC portrait of Col. William A. Anders, USAF

September 1962 MSC portrait of Borman soon after his selection as an astronaut

Borman is among NASA's next nine astronauts (Group 2), posing on the steps of Cullen Auditorium at the University of Houston after a news conference announcing their selection on September 17, 1962. *Front row, left to right*: Lovell, Lt. Charles Conrad Jr., and Borman. *Center row, left to right*: Elliot M. See Jr., Capt. Thomas P. Stafford, and Capt. Edward H. White II. *Back row, left to right*: Neil A. Armstrong, Lt. Cmdr. John W. Young, and Capt. James A. McDivitt. White covers a cut on Borman's head, received while he was checking out an aircraft.

Borman poses with a mockup of a Gemini spacecraft at McDonnell Aircraft Corp. on December 12, 1962. The Group 2 astronauts are at the St. Louis, Missouri, plant for Gemini familiarization by Group 1 astronaut Gus Grissom.

Group 1 and 2 astronauts watch a hammock go up during training from June 3 to 6, 1963, at the Caribbean Air Command's Tropic Survival School at Albrook Air Force Base (AFB) in the Panama Canal Zone. *Left to right*: Armstrong, Conrad, Alan Shepard, Borman, Gordon Cooper, Lovell, John Glenn, See, Gus Grissom, and Stafford. At right is MSC pilot and instructor Bud Ream.

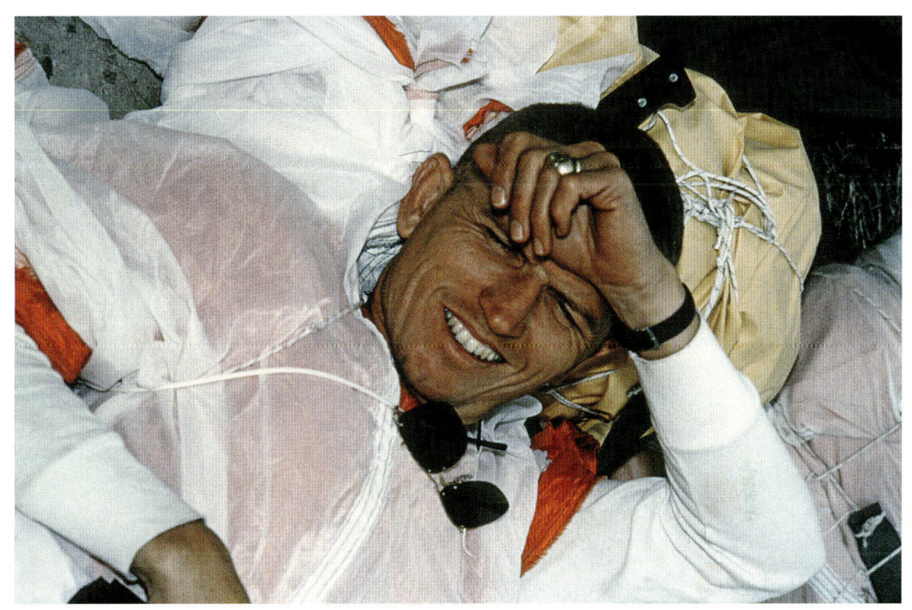

Borman participates in desert survival training at Carson Sink north of Stead AFB in western Nevada from August 5 to 10, 1963. Stead was home to the USAF's Survival, Evasion, Resistance, and Escape School.

Group 2 astronauts pose in garb fashioned from parachutes during the training at Carson Sink. *Left to right*: Armstrong, Borman, Conrad, Lovell, McDivitt, See, Stafford, White, Young, and Deke Slayton, director of flight crew operations. *Bill Taub archive*

Borman stands inside the M-1 mockup on September 17, wearing a prototype pressure suit. The review approves having the Lunar Excursion Module (LEM) crewmen stand rather than sit, a concept MSC had formally proposed in August. *Bill Taub archive*

Borman prepares for parasailing training at Ellington AFB, southeast of Houston, on September 1, 1963. Attached to a parachute, he'd become airborne after a few strides as he is towed aloft by a 600-foot line behind a truck moving at 35 mph. *Bill Taub archive*

Left to right: Borman, Slayton, and Young at Grumman Aircraft Engineering Corp. in Bethpage, New York, on September 17, 1963, during a NASA-Grumman review of the LEM mockup M-1 from September 16 to 18. *Bill Taub archive*

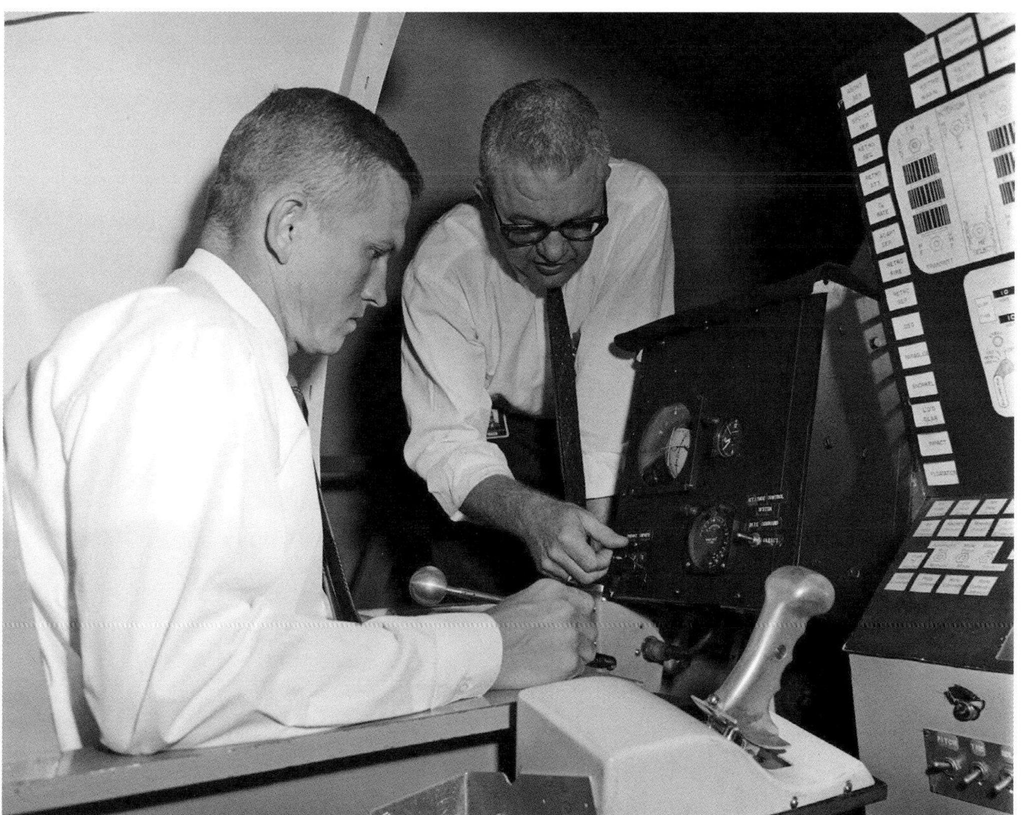

Armstrong (*left*) chats with Borman during training at the Water Safety and Survival School at Naval Air Station Pensacola, Florida, on September 24, 1963. *Bill Taub archive*

Borman at the school on September 24 during early Gemini training for his group. *Bill Taub archive*

An engineer demonstrates the Visual Docking Simulator to Borman (*left*) at NASA's Langley Research Center in Hampton, Virginia, on November 1, 1963. The system uses a television camera, mirrors, and a computer to create the illusion that Gemini astronauts are guiding the spacecraft toward an Agena docking target.

Borman (*right*) during a geologic excursion to the Grand Canyon, Arizona from March 12 to 13, 1964. Astronauts Wally Schirra and White are at left, with Al Chidester of the US Geological Survey at center.

Borman (*second from right*) serves as backup command pilot for Gemini IV with backup pilot Lovell (*right*). The prime crew of White (*left*) and McDivitt pose at MSC with a Gemini model in its skid-landing configuration on July 29, 1964.

Borman prepares for launch at Cape Kennedy Air Force Station (AFS), Florida, as command pilot of Gemini VII on December 4, 1965.

The erector is lowered at Launch Complex (LC)-19 before the Gemini VII launch on December 4.

Borman during a medical test aboard Gemini VII. (16 mm still frame)

Gemini VII is photographed by the crew of Gemini VI-A on December 15, 1965, during their rendezvous in Earth orbit, which is a revised mission after Gemini VI's Agena target had exploded after launch six weeks earlier.

Lovell's first astronaut portrait at MSC in late 1962

The official Group 2 portrait by *Life* magazine photographer Ralph Morse in September 1963. Seated around a Gemini model are (*clockwise from top right*) Borman, Young, Stafford, Conrad, McDivitt, Lovell, See, White, and Armstrong.

Lovell poses in front of a Gemini mockup during the Group 2 familiarization visit at McDonnell on December 12, 1962. An Agena mockup is at right.

Lovell (*front row, third from left*) is among the astronauts undergoing tropical survival training from June 3 to 6, 1963, near Albrook AFB in the Panama Canal Zone. *Front row, left to right*: Conrad, Shepard, Lovell, and Armstrong. *Back row, left to right*: Grissom, instructor Ray Zedekar, Stafford, See, Glenn, Cooper, MSC's Bud Ream, and Scott Carpenter.

Lovell (*center*) joins Young, Slayton, and See to practice knot tying at Carson Sink, Nevada from August 5 to 10, 1963.

Lovell during the Nevada desert training

Lovell (*left*) and See at Grumman on September 17, 1963, during the M-1 mockup review; M-1 is in the background. *Bill Taub archive*

Lovell during water survival school training on September 24, 1963, at Naval Air Station Pensacola. *Bill Taub archive*

Lovell dons a Gemini pressure suit at the school on September 24. In the background are Conrad, Armstrong, and Borman. *Bill Taub archive*

Lovell (*left*) in the Visual Docking Simulator at Langley Research Center on November 1, 1963

Lovell and Stafford participate in geology training west of Big Bend National Park near Marathon, Texas, from April 15 to 16, 1964.

Lovell (*right*) and Borman (*left*) flank the Gemini 4 prime crew of McDivitt and White during a visit to the US Capitol in Washington, DC, on April 29, 1965, to meet members of Congress. The four also take part in a news conference at NASA headquarters.

Lovell wears a lightweight G5C pressure suit, used only on this mission, before his Gemini VII launch on December 4, 1965, at Cape Kennedy AFS.

Gemini VII lifts off from LC-19 atop its Titan II booster at 2:30 p.m. EST on December 4.

Lovell gazes out his window aboard Gemini VII during the two-week mission, which set the record for the longest space flight until Soyuz 9 in 1970. (16 mm still frame)

Gemini VII is photographed from Gemini VI-A on December 15, 1965.

Lovell (*left*) will be command pilot for Gemini XII, the final mission of the program. His pilot will be Buzz Aldrin (*to his left*). They're joined by backup crewmen Gordon Cooper and Gene Cernan (*right*) at McDonnell in September 1966.

Aldrin (*left*) and Lovell aboard Gemini XII in November 1966

Aldrin (*left*) and Lovell are welcomed aboard prime recovery ship USS *Wasp* on November 16, 1966, by RAdm. Percival Jackson, commander of Carrier Division 14.

First NASA portrait of Anders in 1963

Anders is among fourteen new astronauts (Group 3) posing outside the University of Houston's Cullen Auditorium after the announcement of their selection on October 18, 1963. *Left row, top to bottom*: Capt. R. Walter Cunningham, USMC (Res.); Maj. Edwin E. "Buzz" Aldrin Jr., USAF; Capt. Michael Collins, USAF; Capt. William A. Anders, USAF; Lt. Alan L. Bean, USN; and Lt. Eugene A. Cernan, USN. *Center row, top to bottom*: Lt. Cmdr. Richard F. Gordon Jr., USN; Capt. Clifton C. Williams Jr., USMC; Capt. Charles A. Bassett II, USAF; Capt. Russell L. "Rusty" Schweickart, ANG; and Lt. Cmdr. Roger B. Chaffee, USN. *Right row, top to bottom*: Capt. Theodore C. Freeman, USAF; Capt. Donn F. Eisele, USAF; and Capt. David R. Scott, USAF.

Scott (*left*) and Anders visit LC-37 at Cape Kennedy AFS on April 9, 1964, with Saturn/Apollo (SA)-6 on the pad. At right is Charles Friedlander, chief of the Astronaut Support Office at the Kennedy Space Center (KSC).

Anders observes an image of the sun projected on a table through the optical system of the world's largest solar telescope at the Kitt Peak National Observatory, near Tucson, Arizona. A group of seventeen astronauts visit the observatory as part of a three-day geology field trip to Flagstaff from April 30 to May 2, 1964.

Bean, Anders, unidentified, Cunningham, Hal Masursky of the US Geological Survey, and Schweickart examine the sun's projected image at the Kitt Peak National Observatory.

Left to right: Freeman, Anders, and Bean during geology training at the Philmont Boy Scout Ranch in Cimarron, New Mexico, on June 3, 1964

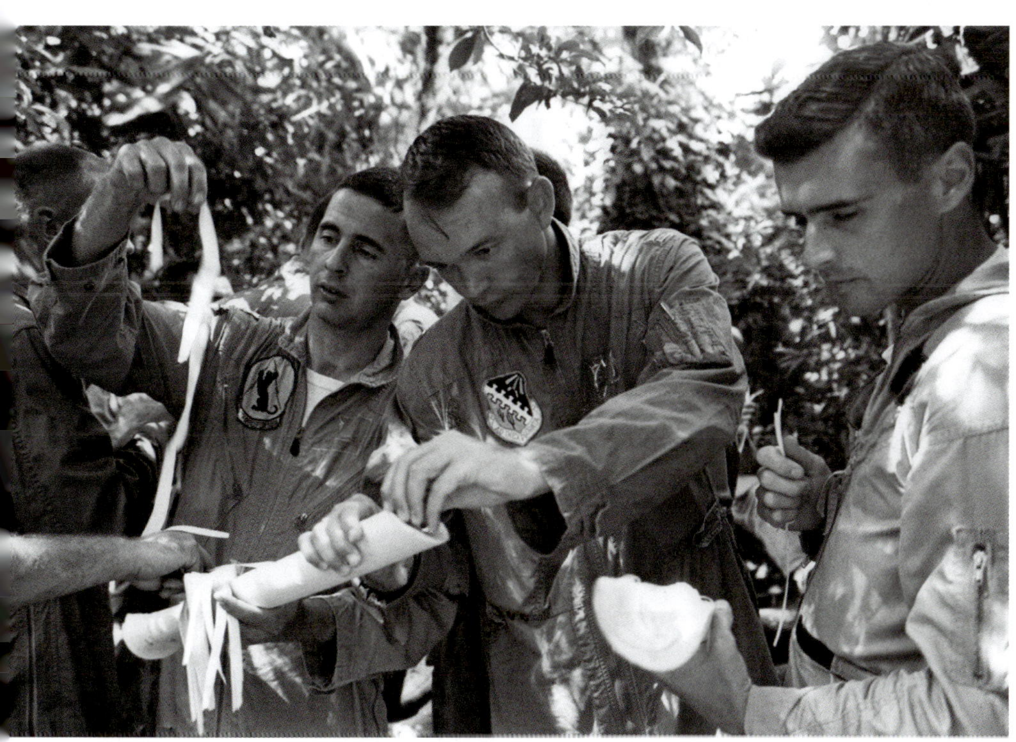

Anders (*left*), Collins, and Chaffee sample heart of palm during jungle training at the Tropical Survival School in the Panama Canal Zone during the week of June 22, 1964.

Anders examines a horned toad during desert survival training at Carson Sink, Nevada, from August 10 to 14, 1964.

Anders (*left*) and Bassett inspect samples during a geological field trip with other astronauts and geologists to Drekagil, Iceland, from July 9 to 18, 1965.

Schweickart (*left*) and Anders examine a rock during the trip to the Drekagil ravine in northern Iceland.

Anders makes a new friend during the Iceland visit.

Anders (*left*) goes over Gemini XI pilot Dick Gordon's environmental control unit, required for his planned spacewalk during the mission to dock with an Agena target vehicle. They work at the boresight range (the "timber tower") at KSC on July 24, 1966. The facility tests telemetry and radar capabilities for docking. Anders is Gordon's backup.

CHAPTER 2

Launch Complex 39

A Saturn V launches from Launch Complex (LC)-39 at KSC in this April 1963 illustration by NASA artist Don Mackey. A second Saturn V rolls out (*left*) from the Vehicle Assembly Building (VAB), known initially as the Vertical Assembly Building, as a Saturn stage arrives at the turning basin (*right*).

A December 1963 illustration of the VAB at sunset, with a Saturn V on a mobile launcher (ML) rolling out at right and a crawler-transporter emerging at left.

Launching astronauts to the Moon required the largest rocket ever built at the time, the Saturn V, which in turn required constructing a sprawling new facility at the north end of Merritt Island, Florida, known as Launch Complex 39. Instead of being assembled at the launchpad, the Saturn V stages and Apollo spacecraft would be stacked on one of three mobile platforms and checked out first inside what was the world's largest building, the VAB. One of two huge crawlers would then transport the assembled rocket and Apollo spacecraft more than 3 miles to one of two launchpads. The new complex, built between 1963 and 1966, led some to dub KSC a "Moonport."

The final design of the hardstand for LC-39's Pad A in June 1964

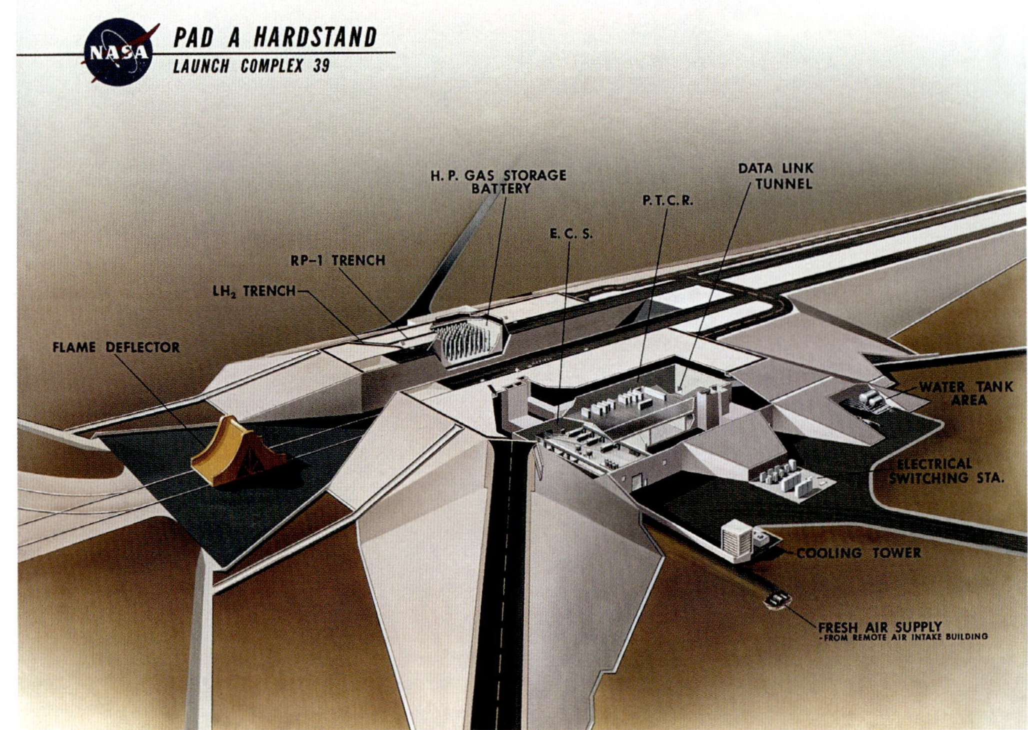

Pad A begins to take shape on the north end of Merritt Island in January 1963. The 80-foot pile is made of shell and sand dredged from the Banana River by the Gahagan Dredging Corp. of Tampa, Florida.

The path of the crawlerway to the VAB begins to emerge as seen on June 6, 1963. Bulldozers, draglines, and other earth-moving equipment mold the pile into shape.

Hardstand construction continues in July 1963, with the Atlantic Ocean in the background. The dredging operation pumps almost 7 million cubic feet of material from the river; much of it is used for the crawlerway.

Just more than 3 miles to the west-southwest, some 4,200 steel pipes are driven to bedrock through sandy soil to support the VAB, photographed on September 27, 1963. *At upper right*, concrete forms are in place for the foundation. Eventually, almost 42,000 cubic yards of concrete are poured for pile caps and the floor slab before the foundation is completed in May 1964.

Apollo contractor executives visit the VAB construction site on December 3, 1963, as pile drivers continue to pound pipes into place.

Construction of the crawlerway, formed from dredged river material, is underway in January 1964. It has two 40-foot-wide lanes, separated by a 50-foot median. It would be filled with more than 3 feet of subbase material and topped by 3 feet of graded crushed aggregate along a roadbed 130 feet wide.

The octagonal shape of Pad A begins to take shape on January 14, 1964.

An ML base is under construction on February 7, 1964, near one of six huge support stands it will rest on.

The framework for a launch umbilical tower (LUT) rises from an ML on May 11, 1964.

Lt. Col. Rocco Petrone uses models to brief members of the Aviation/Space Writers Association near the VAB on the afternoon of May 28, 1964. Petrone has been detailed from the US Army to serve as Saturn project officer.

The steel skeleton of the VAB rises on July 8, 1964, with the four-story Launch Control Center (LCC) under construction in the foreground. US Steel's American Bridge Division fabricated and erected the structure.

One of the "trucks" that will be mounted at the four corners of a huge crawler-transporter for LC-39 undergoes tests at the Marion Power Shovel plant in Marion, Ohio, on July 15, 1964. Each of the two tracks has fifty-seven treads and moves by four diesel-powered electric motors.

Crawler-transporter construction proceeds at KSC in October 1964, with treads in the foreground.

Two of this crawler's four trucks have been added by November 9, 1964; yellow treads are in four groups at lower right. The two crawlers will be the largest self-powered vehicles in the world.

The VAB's low bay area (*left*) and the LCC (*right*) take shape on November 9. Behind the high bay skeleton are three LUTs under construction atop their MLs.

The flame trench at Pad A, 59 feet wide and 450 feet long, is under construction in January 1965. The two towers will provide pad access to the top of the ML and the LUT's elevator.

This view of VAB construction looks east toward Pad A on January 7, 1965. The 12-mile canal for barges bringing Saturn stages to the VAB, and the turning basin for them, *at right*, have been dredged by the Gahagan Dredging Corp. The crawlerway is topped with an asphalt sealer and 8 inches of smooth Alabama river rock to reduce friction.

J. Robert Oppenheimer (*center*), director of the Institute for Advanced Studies in Princeton, New Jersey, tours KSC on January 20, 1965. *At left*, Chuck Friedlander of NASA Public Affairs, with, *at right*, Dr. Adolph Knothe, KSC chief scientist. Oppenheimer spoke the previous evening at a seminar at Patrick AFB sponsored by Brevard Engineering College.

Aerial view of the VAB in January 1965, with work platforms under construction at the bottom of the photo

VAB construction on January 22, 1965, with two LUTs at left

Petrone (*pointing*) discusses VAB and LC-39 construction with *Today* host Hugh Downs on February 6, 1965, for a segment on the NBC-TV morning show about the upcoming first manned Gemini launch.

Ingalls Iron Works personnel watch as a crane tops off ML-3 on March 3, 1965.

An Ingalls worker near ML-3. The Birmingham, Alabama, firm had begun construction of the three LUTs in July 1963, each 380 feet tall.

An Ingalls worker poses with LUT-3 in the background.

The VAB is framed by two nearly complete MLs on April 9, 1965. Each is topped by a yellow hammerhead crane. Service arms have yet to be installed.

On April 14, 1965, Petrone signs the last major steel beam for the VAB, already autographed by NASA, contractor, and Army Corps of Engineers workers, which will be raised to be installed during a topping-off ceremony that morning.

KSC director Kurt Debus speaks at the topping-off ceremony at 11:00 a.m. EST, with a crawler at right. "This building is not a monument," he says. "It is a tool, if you will, capable of accommodating heavy launch vehicles. So, if people are impressed by its bigness, they should be mindful that bigness in this case is a factor of the rocket-powered transportation systems necessary to provide the United States with a broad capability to do whatever is the national purpose in outer space."

The autographed 38-foot beam is secured in place during the ceremony, more than 520 feet up.

Petrone gives producer Walt Disney a hands-on look at a crawler-transporter after the ceremony. Disney had just toured MSC and the Marshall Space Flight Center (MSFC) in Huntsville, Alabama, to research both an update of Tomorrowland at Disneyland and a possible TV movie on Apollo.

The four high bays are empty in this view of the west side of VAB on July 12, 1965.

Construction of the crawlerway ramp to the Pad A hardstand continues on August 5, 1965.

Extensible work platforms up to three stories tall are assembled outside the VAB, because of their size, in September 1965. Each high bay will receive five pairs of the vertically adjustable platforms.

The red work platforms have been installed in the VAB's high bays, and two yellow bridge cranes have a yellow bridge crane has been installed over the transfer aisle, as seen on October 29, 1965.

Outside, construction on an ML continues on October 29.

Pad A nears completion on November 2, 1965, with its liquid oxygen (LOX) tank at upper left. Pad B is under construction in the distance in this view looking north.

Firing Room 1 on the third floor of the LCC is filled with empty racks for control consoles in January 1966. This view looks toward Area C in the back of the room; the large projection screens are above this area. These consoles will be manned by contractor and NASA engineers who monitor various functions of the spacecraft, launch vehicle, and ground support equipment.

This view looks in the opposite direction toward Areas A and B, and the windows facing the pads. The vertical black louvers are closed in the center.

The five pairs of work platforms are in place in high bay 1 in January.

A top-down view of an ML under construction in a photo taken from its crane in January. The square opening provides an outlet for flames and exhaust from the Saturn's first stage at ignition.

The 402-foot-tall mobile service structure (MSS) is under construction on January 12. It will provide work platforms at the pad, and limited weather protection for the launch vehicle. The MSS will be moved to and from the pads by the crawler.

Huge blowers and associated ductwork within each ML, seen in January, provide environmental and thermal control to the Saturn V, including gaseous purging and pressurization.

Sable palmettos frame the east side of the VAB on January 12 as it nears completion. It is the largest building in the world by volume at the time.

Crawler operators in the cab slowly approach ML-1 on January 27, before an important test for the transporter the next day.

The crawler is parked near ML-1 on January 27.

This view from one LUT looks south toward the VAB, with a second LUT at left, on January 28. The Saturn Causeway, the main vehicle access road to and from the pads, runs alongside the crawlerway.

A view of LUT-2 under construction as photographed from LUT-3 on January 28, with the Kennedy Parkway North running left to right in the background

This view looks up at an LUT from under an ML on January 28.

A crawler's generator belches diesel exhaust as it lifts ML-1 for a move into the VAB for the first time on January 28.

Personnel inside Firing Room 1 in the LCC watch as the ML moves along the crawlerway at about 10 feet a minute toward the VAB.

The crawler and ML slowly pass the LCC.

The ML nears the VAB, with the LCC at left.

View of the LUT and ML from above high bay I

ML-1 carefully enters high bay 1, where a complete "dummy" Saturn V will be stacked for facility fit checks.

On March 4, ML-3 passes the MSS (under construction) as it heads toward Pad A for the first time.

ML-3 arrives at the hardstand at Pad A later that day.

LC-39 engineering manager Don Buchanan (*left*) receives a cigar from Petrone at Pad A on March 4 for a job well done with ML-3.

ML-3 heads back to the VAB at sunset on April 3, with a liquid nitrogen tank at right.

Workers check out service arm no. 1 (S-IC Intertank) at the 60-foot level on ML-1 on March 15, 1966. It provides liquid oxygen (LOX) fill and drain interfaces. The arm would be retracted thirty seconds before launch within thirteen seconds. The arms are built by the Hayes International Corp. of Birmingham, Alabama. For scheduling reasons, NASA has Hayes deliver arms directly to KSC for the first ML; after fit checks, they'd be sent to Marshall for full testing like all other service arms.

A crane slides service arm no. 4 (S-II Intermediate) from the Super Guppy onto a scissor-lift trailer to be towed to the VAB after arrival at the Cape Kennedy AFS Skid Strip on April 15, 1966.

Service arm no. 9 (CM Access Arm) is in the VAB transfer aisle on May 23, 1966

To make sure all the facilities at LC-39 would properly interface with the Saturn V, KSC put together a complete "dummy" launch vehicle (SA-500F). Also known as the Facilities Integration Vehicle, it duplicated a real Saturn V to test stacking the three stages in the VAB, the fit of the service platforms in the high bay, ML-crawler operations, the propellant-loading system, and connections to ML and ground support equipment. The S-IVB stage had been used for fit checks at LC-34 and LC-37. SA-500F was stacked on ML-1 up to the instrument unit (IU) in high bay 1 on March 30, 1966. The Apollo CSM facilities verification boilerplate was added on May 2, and the stack was rolled out to Pad A on May 25.

The first stage of SA-500F, S-IC-F, is moved into position over ML-1 on March 15, 1966. Only one F-1 engine has been installed for the pad tests.

The second (S-II-F) and third (S-IVB-F) stages of SA-500F are stacked by April 1, with the high bay service platforms retracted.

The CM's mockup launch escape system (LES) arrives in the VAB transfer aisle on May 4.

Two glassed-in areas in the front corners of Firing Room I in Area A are the Operations Management Room for senior NASA officials (*left*) and a visitors' gallery (*right*). Work in the LCC nears completion during May. Senior launch managers (including Debus and Petrone) will sit in the top row; test directors will occupy the second row.

Maj. Gen. Sam Phillips, Apollo program manager (*left*), Petrone, and von Braun share congratulations on May 25, 1966, at the rollout ceremony for SA-500F. *Bill Taub archive*

NASA associate administrator for space flight George Mueller (*left*) and Marshall Space Flight Center (MSFC) director Wernher von Braun at the rollout of SA-500F on the morning of May 25, the fifth anniversary of President John Kennedy's speech setting a deadline to land astronauts on the Moon. "We are well on our way to keeping that commitment," Mueller says at the ceremony. "The Apollo is on schedule." *Bill Taub archive*

Almost nine hours later, SA-500F approaches the Pad A ramp at the intersection with the Saturn Causeway.

SA-500F and its ML are on hardstand supports at sunset photographed through a star filter.

Top-down view of SA-500F on June 2. Six days later, it would be rolled back to the VAB for forty-eight hours as Hurricane Alma approached.

SA-500F at Pad A in July. Daily public bus tours of KSC begin on July 15, which include a view of the rocket and a stop at the VAB.

SA-500F with the MSS in place at right on July 22, 1966. The facilities checkout culminates with filling its fuel tanks with propellant, and the vehicle is returned to the VAB on October 14.

CHAPTER 3

December 1966–December 1967

Left to right: Astronauts Bill Anders, Frank Borman, and Mike Collins visit NASA's Michoud Assembly Facility in New Orleans, Louisiana, on December 13, 1966. On December 22 they would be selected as the Apollo 3 (AS-503) crew, planned to be the first manned Saturn V launch, including an LM, into Earth orbit. Behind them is S-IC-3, the first stage of that launch vehicle, with its five F-1 engines manufactured by Rocketdyne, an NAA division in Canoga Park, California.

Anders (*hard hat, left*) and Borman are among five astronauts at Michoud for a briefing on the Saturn V. Behind Borman with glasses is John Cully, S-IC project manager for Boeing, the stage's prime contractor.

Borman and Collins (*second from right*) in a Michoud control room

Borman climbs from the cockpit of a Lockheed T-33 Shooting Star jet after arriving at Langley Research Center in Hampton, Virginia, for a facilities visit in January 1967.

The Michoud Assembly Facility is an 832-acre NASA manufacturing complex in New Orleans East, a section of New Orleans, Louisiana. First created in 1940 for World War II Higgins boat production, Michoud came under NASA management in 1961 as part of the Marshall Space Flight Center in Huntsville, Alabama. From September 1961 to December 1972, the site was utilized first by the Chrysler Corp. to build the first stages of the Saturn I and Saturn IB, and later by the Boeing Corp. to build the first stage of the Saturn V. From 1973 to 2010, the factory was used by the Martin Marietta Corp. to manufacture the space shuttle's external fuel tanks. One of the largest manufacturing plants in the world, it has 43 environmentally controlled acres (1,870,000 sq. ft.) under one roof.

Anders (*left*) and Collins arrive at Langley in a Northrop T-38 Talon trainer.

Anders, Borman, Collins, and a Langley engineer enter the Flight Research Laboratory (building 1244), the hangar housing the Apollo Rendezvous Docking Simulator.

The group pauses in front of the docked LM and CM mockups, which can be suspended and moved on overhead cranes. The system, built for Gemini, has been reconfigured for Apollo.

Borman (*in flight suit and hard hat at center*) and backup crew LM pilot C. C. Williams (*left of Borman*) pose with personnel at the Lunar Landing Research Facility at Langley Research Center. Astronauts could simulate lunar landings by using the mock LM behind them, powered by a small rocket motor while suspended from a crane.

Test stand Beta III at the Douglas Aircraft Corp.'s test facility east of Sacramento, California, shows damage after the S-IVB third stage for AS-503 exploded on January 20, 1967. The explosion happens several minutes before the planned ignition of its J-2 engine.

No one is injured in the explosion, which produces a fireball and shakes windows 10 miles away. The cause is traced to faulty welds in one of eight helium storage spheres. The stage is replaced by S-IVB-504, which is redesignated S-IVB-503N.

A five-member review board chaired by KSC director Kurt Debus begins an investigation, but officials aren't sure whether the accident would delay the Apollo program.

The Apollo I CM at LC-34 shows damage after the January 27 fire that took the lives Apollo I astronauts Gus Grissom, Ed White, and Roger Chaffee. The crews for Apollo 2 and 3, which had been named a month earlier, would be disbanded.

Left to right: Borman, Jim McDivitt, Deke Slayton, Wally Schirra, and Alan Shepard testify on April 17 before the House Subcommittee on NASA Oversight investigating the Apollo I fire in the Rayburn Building in Washington, DC. *Photo by Bob Schutz / AP*

The J-2 engine for what would be Apollo 8's S-IVB (503N) is in the foreground in the Douglas Aircraft's Vehicle Checkout Laboratory near Sacramento in August 1967. Red spheres will contain gaseous helium for the pneumatic system controlling the engine's fuel valves. Behind the engine is stage 207 (used for Skylab 3) on a birdcage (*left*) and stage 206 (used for Skylab 2) in Tower No. 2.

Although no longer officially a flight crew, the astronauts maintain their training proficiency as a team. This double exposure shows Anders, Collins, and Borman entering, and inside, the Apollo Mission Simulator (AMS) in Building 5 at MSC in Houston in August 1967.

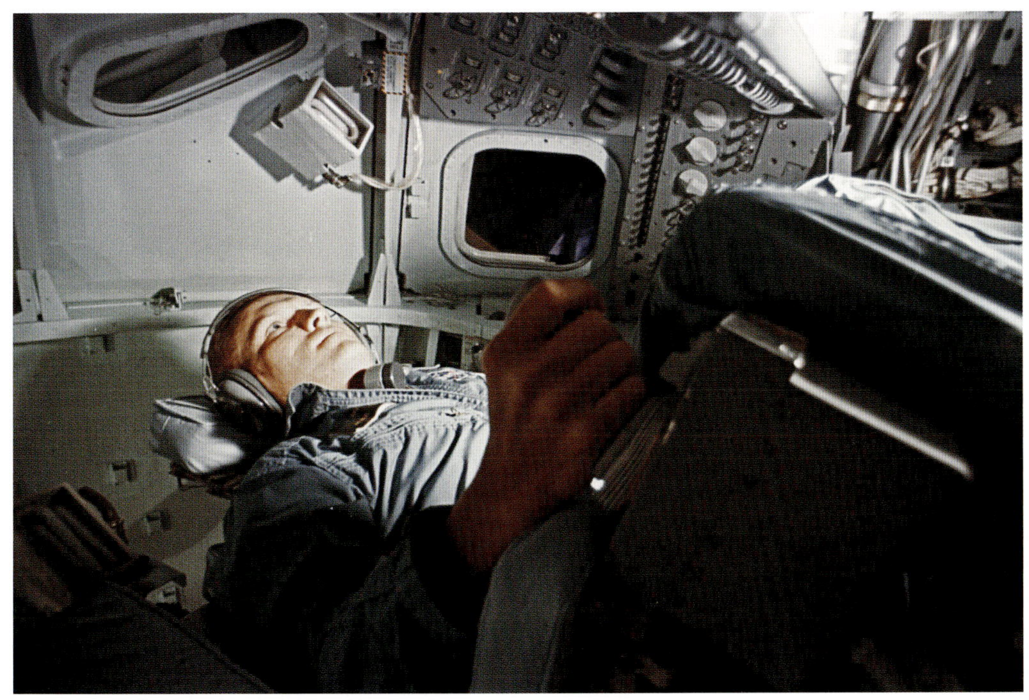

Borman trains in the left-hand couch in the fixed-base AMS. By this time, the Borman crew is tentatively assigned to AS-505, a planned third manned Apollo flight and the second manned Saturn V launch.

Anders in the right-hand couch in the AMS

Collins sights through the scanning telescope at the navigation station in the AMS. A twin simulator is available at KSC.

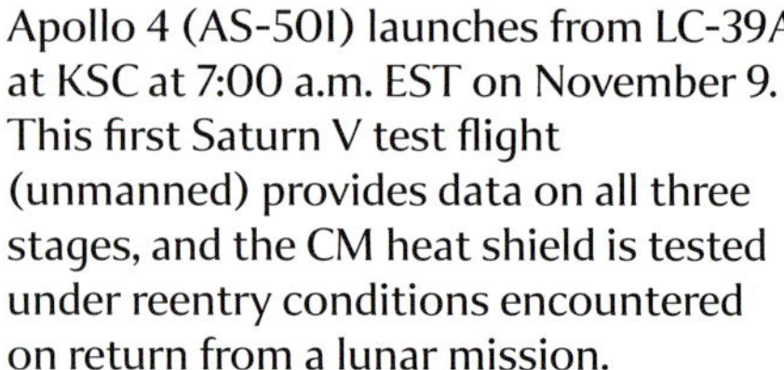

Apollo 4 (AS-501) launches from LC-39A at KSC at 7:00 a.m. EST on November 9. This first Saturn V test flight (unmanned) provides data on all three stages, and the CM heat shield is tested under reentry conditions encountered on return from a lunar mission.

The Apollo 4 CM (CM-017) is hoisted aboard prime recovery ship USS *Bennington*. The spacecraft splashed down at 3:37 p.m. EST, northwest of Hawaii.

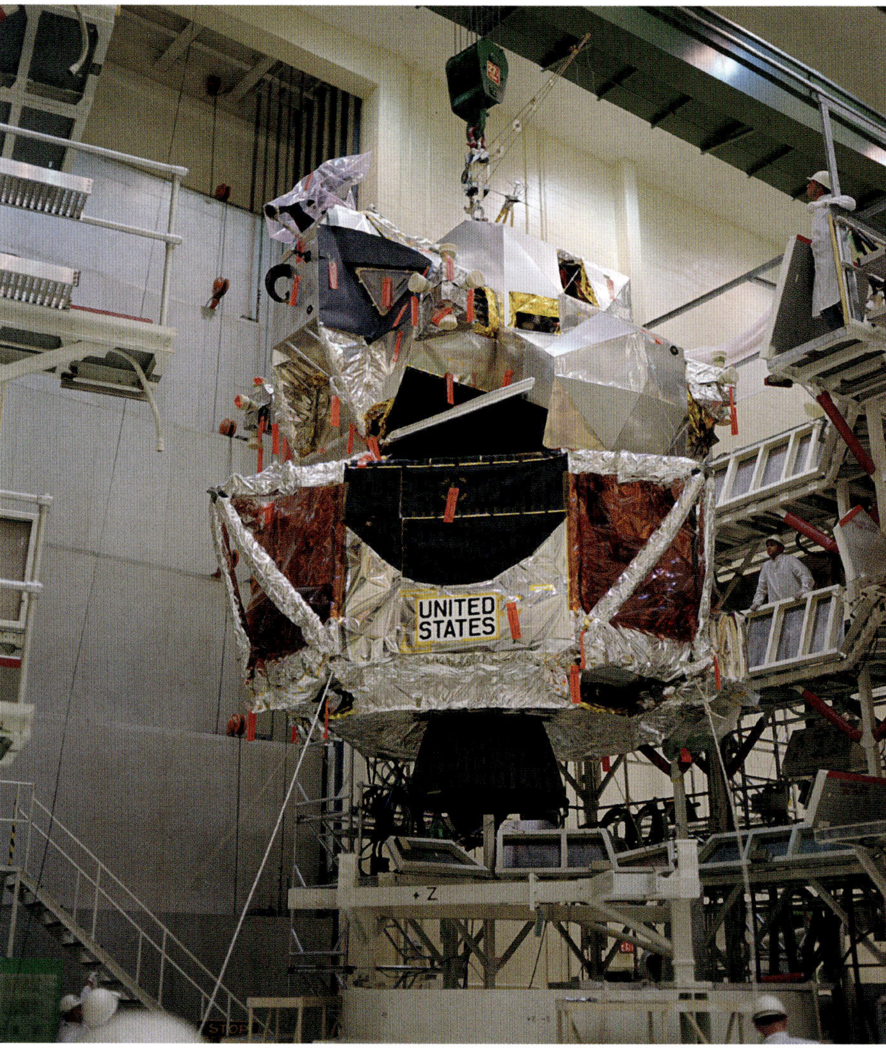

On November 10, LM-1 is raised to be installed in Spacecraft Lunar Module Adapter (SLA)-7 in the Manned Spacecraft Operations Building (MSOB) at KSC. The LM does not have landing legs, to save weight. It is scheduled to be flown on the unmanned Apollo 5 mission in about six weeks.

Nine days later, the LM and SLA are lifted for mating with SA-204 at LC-37.

The first Saturn V flight booster to be assembled at NASA's Michoud Assembly Facility by Boeing is rolled aboard the NASA barge *Poseidon* at Michoud on December 21 for transport to KSC. The four-day trip, which begins December 23, takes the 138-foot-long S-IC-3 first stage across the Gulf of Mexico and around the tip of Florida. Boeing and NASA personnel accompany the stage during its 900-mile voyage. *Poseidon* was built for the US Navy in 1945 as a work barge. It was declared surplus and acquired by NASA. It was converted and named by Avondale Shipyards west of Michoud in 1966.

The second stage of AS-503 (S-II-3) is towed through the Canaveral Barge Canal, heading west from Port Canaveral toward the Banana River, aboard USNS *Point Barrow* on December 24. A USAF pusher ship follows. It has also arrived from Michoud. A one-of-a-kind cargo ship, *Point Barrow* was modified for NASA use in 1965 after Arctic naval operations.

The ships pass through the Canaveral Lock, the largest navigation lock in Florida, built in 1965 by the US Army Corps of Engineers to allow passage of large vessels between Port Canaveral and the Banana River.

The barge arrives at the turning basin at LC-39.

On December 26, 1967, after the Christmas holiday, the second stage is ready to be towed on its carrier to the VAB.

The next day, the first stage (S-IC-3) arrives at the turning basin on *Poseidon*.

Two workers watch as the first stage is lifted by bridge cranes in the VAB's transfer aisle on December 30; the 500F facilities verification CSM mockup is in the background.

The stage's engine fairings and fins have not yet been mounted.

The five Rocketdyne F-1 main engines are protected by covers. Their nozzle extensions will be installed after the stage has been stacked.

A Bendix technician stands near a steadying tagline.

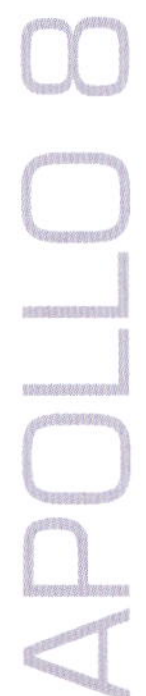

A crane moves the stage into high bay 1.

The S-IC is lowered.

The stage nears ML-1.

The S-IC is slowly guided onto the ML.

S-IC is nearly seated on the ML.

Boeing technicians inspect one of four hold-down arms supporting the stage.

The S-IC is secured on the ML on December 30.

CHAPTER 4

January–July 1968

The instrument unit (IU) for AS-503, the Saturn V that would eventually be Apollo 8, is on its transport pallet inside the Super Guppy on January 4. Nearly 23 feet in diameter, the IU is the Saturn V's brain, responsible for guidance, staging, and telemetry after launch.

Personnel at the Cape Canaveral AFS Skid Strip prepare to unload the IU, which arrives from IBM's facilities at the Marshall Space Flight Center in Huntsville, Alabama.

The IU is moved onto a scissor lift.

A crane lifts the IU and its dolly from the scissor lift.

The IU and its wheeled dolly are lowered to the ground to be towed to a checkout cell in the VAB.

Dusk comes to LC-37 at Cape Canaveral AFS on January 21, 1968, as the unmanned Apollo 5 (AS-204) is prepared for launch the next evening.

Apollo 5 lifts off at 5:48 p.m. EST on January 22, just before sunset. The rocket's S-IVB second stage sends LM-1 into orbit, where its ascent and descent engines, and its ability to abort a landing, are demonstrated. The LM's first test flight is called a complete success.

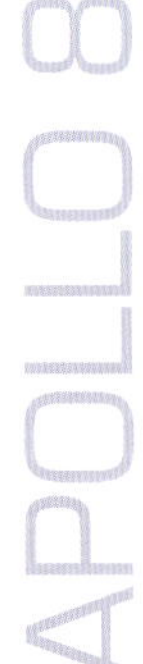

The thrust structure dominates this photo of the first stage (S-IC-3) of AS-503 on its ML in the VAB's high bay 1 in January; it had been stacked about two weeks earlier. The ML's three hooded tail service masts surround the stage. At left is a theodolite, a precision optical instrument for measuring angles in the horizontal and vertical planes. Although scheduled at the time to be an unmanned flight, it is the first stage for what would eventually be Apollo 8.

The S-II stage for AS-503 and its interstage are raised in the VAB transfer aisle for stacking on January 31. The interstage is fabricated at North American Rockwell (NAR)'s Tulsa, Oklahoma, plant. NAR, headquartered in Downey, California, is prime contractor for the CSM and the Saturn V's second stage.

The S-II stage (S-II-3) heads toward high bay 1. It is the most powerful hydrogen-fueled launch vehicle in production.

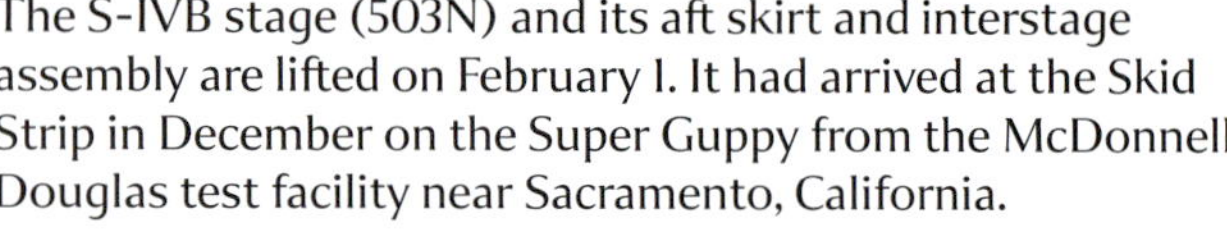

The S-IVB stage (503N) and its aft skirt and interstage assembly are lifted on February 1. It had arrived at the Skid Strip in December on the Super Guppy from the McDonnell Douglas test facility near Sacramento, California.

The S-IVB rises from the transfer pad, the red circle on the floor designating where the Saturn's upper stages should be positioned for lifting.

The S-IVB is lowered in high bay 1, showing its single Rocketdyne J-2 engine nozzle. LUT-1 is behind it.

Technicians inside the forward skirt of the second stage monitor the S-IVB mating on February 1. Two of four S-II retrorockets seen on the aft skirt will pull the second stage away at separation. Service arm no. 5 (S-II Forward) is in place at bottom left. The techs will exit through the hatch at lower left and leave along the arm.

The S-IVB is successfully stacked on February 1, with work platforms surrounding the second stage.

A crane lifts the IU for AS-503 from its carrier for stacking on the same day, with the payload for the flight in the background. It consists of a boilerplate CSM (BP-30) and Spacecraft LM Adapter (SLA-10), which contains a LM test article (LTA-B) to simulate the mass of a real LM.

The IU passes the second stage on its way up.

Technicians inside the S-IVB's forward skirt are ready for IU stacking.

The IU is carefully lowered toward the S-IVB. Service arm no. 7 (S-IVB Forward/IU) is at left.

The next day, February 2, workers in a bucket crane hook BP-30 and SLA to a bridge crane, using the CM's LES attach points.

The pair are hoisted in the transfer aisle for stacking, showing LTA-B in the SLA.

The BP and SLA are lowered toward the IU for mating in high bay 1.

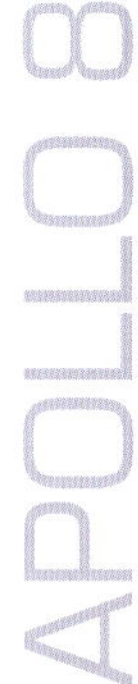

On February 10, astronaut Gordon Cooper (*right*) will assess the effects of simulated launch vibrations on a crewman's vision in CM-105 at the Vibration and Acoustic Test Facility in Building 49 at MSC.

Cooper, in the commander's seat, will be joined by two mannequins. Three cameras will record the test.

Cooper egresses the CM after two brief tests, wearing a prototype headset.

On March 5, Borman, Collins, and Anders take part in a crew compartment stowage review for LM-4 at Grumman Aerospace in Bethpage, New York. They use LM Test Article 1 (LTA-1), configured to resemble the LM-4 flight vehicle. This crew is still training to fly CSM-104/LM-4 in Earth orbit later in 1968. Backup crew members Armstrong, Lovell, and Aldrin also participate.

Borman (*left*) discusses the stowage review with Collins and support astronaut and LM specialist Jerry Carr.

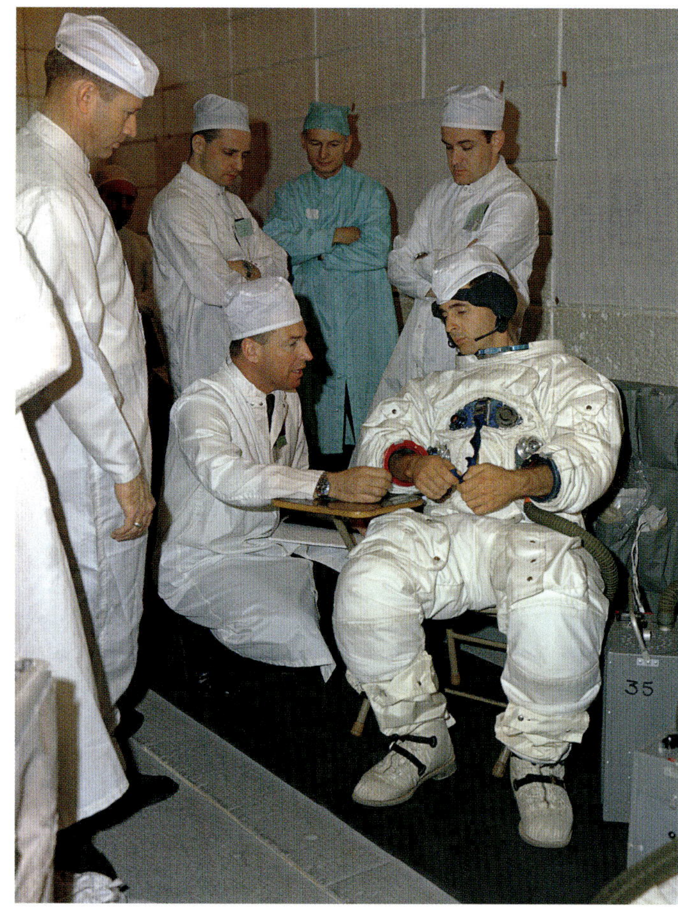

Anders (*suited*) prepares to enter LTA-1 during the stowage review. Backup crewman Jim Lovell kneels beside him, with Carr at left. Veteran MSC suit technician Joe Schmitt is in the aqua smock.

Borman, carrying his suit ventilator, poses with the descent stage of LM-4, still under construction at Grumman.

Anders with the LM-4 descent stage. The crewmen wear clean-room caps over their headsets.

Original Apollo 9 backup commander Neil Armstrong and backup lunar module pilot Buzz Aldrin pose with the LTA-1 ascent stage. Both wear an early version of the Apollo headset.

NAR managers monitor a CM stowage review on closed-circuit TV in Downey, California, on March 15. Participating are AS-504 crewmen McDivitt, Scott, and Schweickart, plus their backups Conrad and Gordon. AS-505 crewmen Borman, Collins, and Anders repeat the test later in the day.

Aldrin sits in the hatch of a CM mockup during the stowage review.

Left to right: Anders, Collins, and Borman at the review

Apollo 6 (AS-502) waits at LC-39A during a lighting test on March 29, with the pad's LOX tank at right. The Mobile Service Structure (MSS) moves away at left.

Apollo 6 heads skyward at 7:00 a.m. EST on April 4, the second unmanned Saturn V test flight. Several problems turn up: Excessive g-forces caused by unexpectedly severe "pogo" oscillations occur in the first stage, two second-stage engine shutdowns leave the CSM in a less-than-nominal elliptical orbit, and the S-IVB engine later fails to restart.

The S-II interstage falls away in this 16 mm motion picture still frame captured by the only one of five cameras on the Saturn that is recovered. The film becomes iconic Apollo footage.

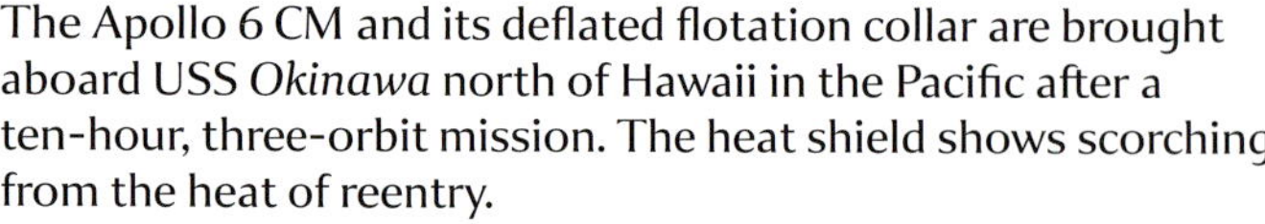

The Apollo 6 CM and its deflated flotation collar are brought aboard USS *Okinawa* north of Hawaii in the Pacific after a ten-hour, three-orbit mission. The heat shield shows scorching from the heat of reentry.

On April 4, CM-007A is lowered into Galveston Bay, Texas, by Motor Vessel (MV) *Retriever* for a forty-eight-hour recovery flotation test. *Retriever* is a WWII–era landing craft, utility (LCU), transferred to NASA from the US Army. It was used to train astronauts for post-splashdown ocean recovery operations during the Gemini and Apollo programs from 1963 to 1972. It operated primarily in Galveston Bay and the nearby Gulf of Mexico. Astronauts Jim Lovell, Charlie Duke, and Stu Roosa are aboard.

Roosa (*foreground*) and Lovell during the flotation test

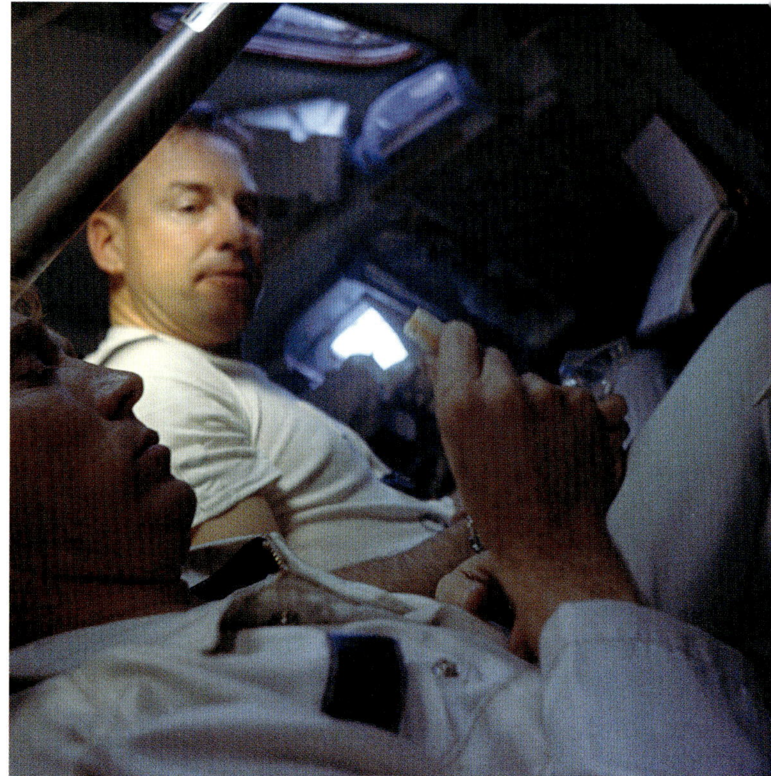

The CM in its stable I (upright) position with its flotation bags inflated on April 4, with *Venus* in the background carrying news photographers.

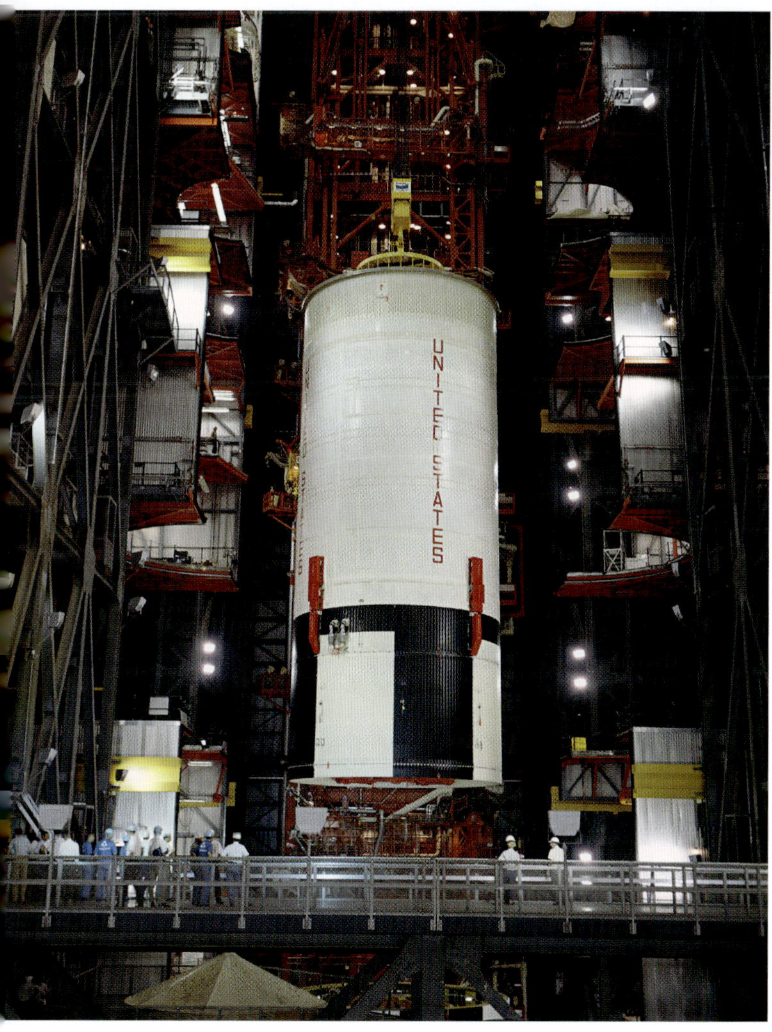

The AS-503 second stage is destacked on April 29 after NASA decides to use it for a manned mission (Apollo 8). It will be returned to the Mississippi Test Facility for modifications to certify it as man-rated. Two of the stage's five liquid hydrogen fuel lines are visible.

The S-II is lowered onto a wheeled workstand, showing its five J-2 engine nozzles, so the interstage can be removed.

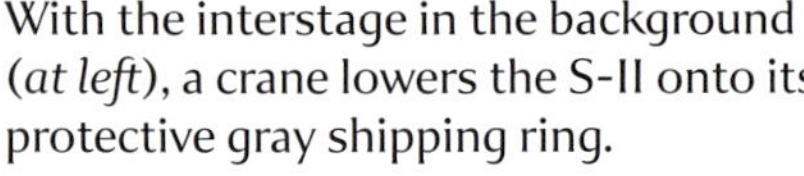

With the interstage in the background (*at left*), a crane lowers the S-II onto its protective gray shipping ring.

The S-II is pushed out of the VAB on April 30.

The stage heads toward USNS *Point Barrow*, a unique cargo ship, on April 30.

The S-II is loaded aboard *Point Barrow* for transport to the Mississippi Test Facility.

After destacking, the SLA for Apollo 8 (SLA-10) is raised in the MSOB on April 30, exposing LTA-B.

The SLA on April 30. On the right is a SLA with SM-102, which had been stacked on the Apollo 7 Saturn IB at LC-34 for testing purposes. It had been destacked several days earlier because of a small fuel spill at the pad.

The SLA is lowered on April 30.

The original Apollo 9 prime and backup astronauts pose with MIT engineers who developed the COLOSSUS software for the CM's guidance and navigation system during a briefing at the MIT Instrumentation Laboratory in Cambridge, Massachusetts. *Standing, left to right*: James Nevins, Albert Hopkins, Robert Bairnsfather, Raymond Morth, Steven Copps, Joseph Turbull, Jerry Levine, Gene Muller, Norman Sears, Fred Martin (team leader), Ivan Johnson, and Russell Larson. *Seated, left to right*: Lovell, Borman, Collins, Anders, Armstrong, and Aldrin.

After modifications, the Apollo 8 second stage (S-II-3) arrives back at KSC on June 27 aboard USNS *Point Barrow*. The liquid hydrogen tank has been qualified for higher flight pressures to become man-rated.

The S-II is soon back in the VAB's transfer aisle, with the S-IVB aft skirt at right.

Below: The next day, the S-II is lowered onto a workstand to be stored in a low bay in the VAB, with S-II-4 for Apollo 9 in the background.

S-II-3 and S-II-4 are on low bay workstands on June 28.

An NAR technician works in the center couch during testing at Downey on June 28.

On July 8 the original Apollo 9 crew poses with NAR workers in Downey during an inspection of CM-104. Borman kneels (*at center*) and Collins stands in the second row (*third from left*), with Anders standing at far right. Support crewman Carr kneels below Anders.

Borman signs autographs for NAR workers at Downey during the inspection. Anders is at right, with his back to the camera.

The crew compartment heat shield for CM-104 is attached to the aft heat shield at NAR on July 22.

The S-II stage is ready for lifting by a bridge crane on July 24.

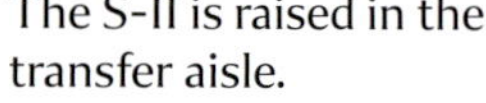

The S-II is raised in the transfer aisle.

The S-II is lowered for mating with the S-IC.

The S-II is stacked in high bay 1. Service arm no. 2 is at lower left.

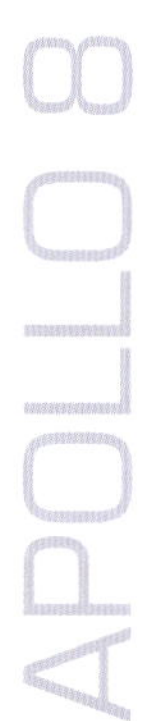

Workers in the VAB get a close view of the stacking.

CHAPTER 5

August–September 1968

CM-103 undergoes final checkout in a test rig in Building 290 at NAR in Downey on August 8. It is wrapped in blue plastic tape, which will not be removed until several days before launch, when the boost protective cover (BPC) is installed at the pad. The tape protects the underlying aluminized Kapton tape covering the CM to provide thermal control.

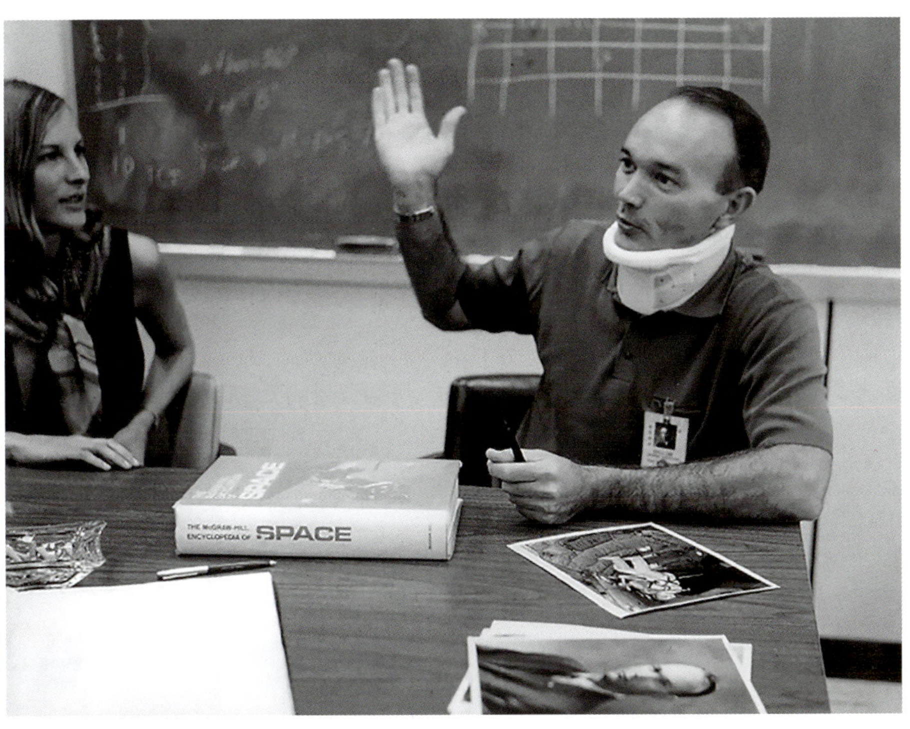

Collins wears a neck brace during an August interview at MSC with French journalist Jacques Tiziou. *Photo by Jacques Tiziou*

Collins with his McDonnell Gemini model. *Photo by Jacques Tiziou*

Collins with his Gemini hard hat. *Photo by Jacques Tiziou*

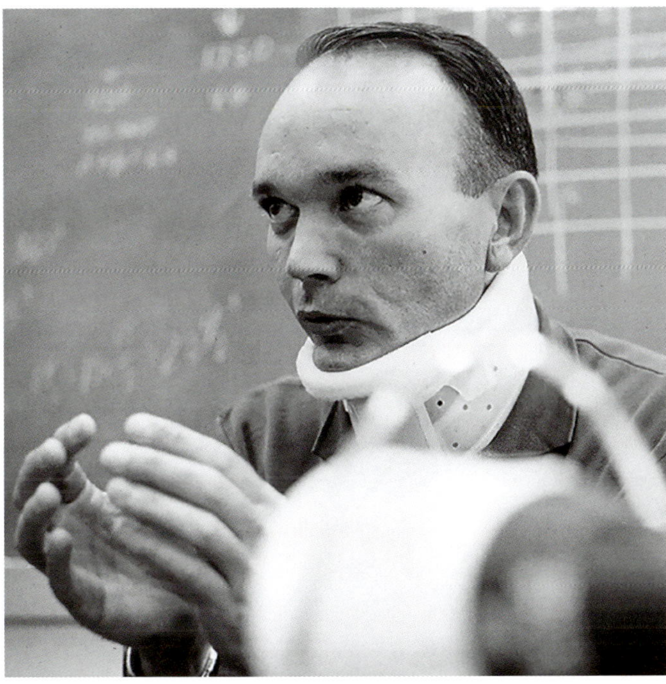

In June 1968, Mike Collins sought treatment for recurring leg problems, diagnosed as a cervical disc herniation. Two vertebrae were fused during surgery at Lackland AFB, Texas, on July 23. His planned recuperation time was three to six months, which Collins spent in a neck brace. As a result, he was removed as CM pilot for AS-504 (Apollo 9) on August 8, and his backup, Jim Lovell, replaced him, joining Borman and Anders. The move created what would be the final Apollo 8 crew.

LM pilot Aldrin was then moved to CM pilot on the Apollo 9 backup crew, support crewman Fred Haise replaced him, and Haise was replaced by Jack Lousma. On August 19, NASA officials announced that LM-3 would not be ready in time for Apollo 8 and would be reassigned to Apollo 9 with the McDivitt crew in Earth orbit. The revised Borman crew was then assigned to Apollo 8, which Lt. Gen. Sam Phillips, Apollo program director, portrayed primarily as a manned Saturn V test, adding he didn't consider a moon-orbiting flight probable.

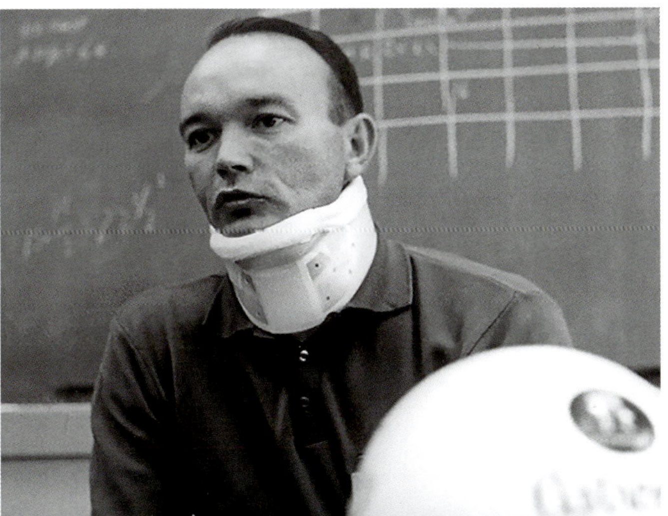

On August 11, service module (SM) no. 103 is unloaded from the Pregnant Guppy aircraft at the Cape Kennedy AFS Skid Strip after arriving from NAR in Downey. The dice on the front sticker show an eight.

The SM is next uncrated in the MSOB. Insulation covers the tops of the cryogenic storage tanks.

The aft view shows the covered mounting ring for the SPS nozzle. "NON FLIGHT HARDWARE" steamers hang from several locations.

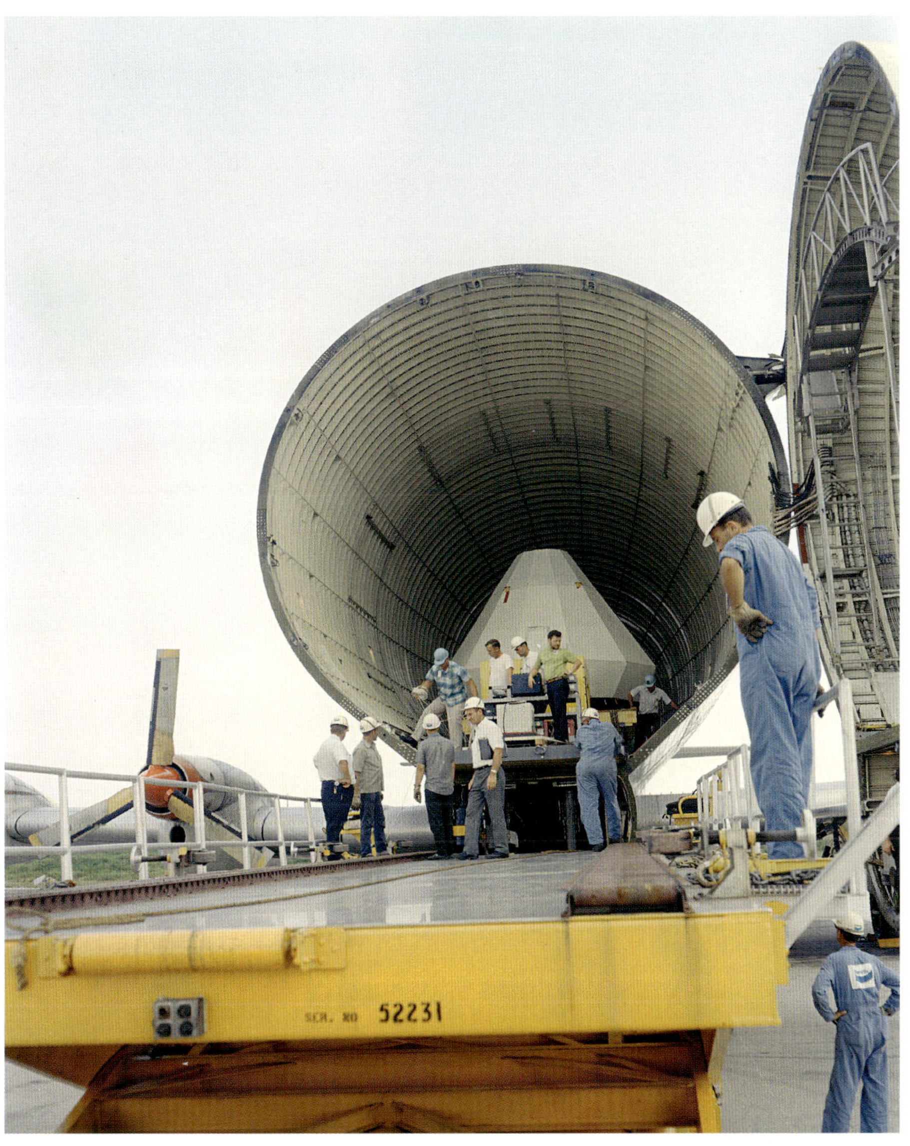

The next day, Bendix personnel raise a scissor lift to accept CM-103 from the Super Guppy after it arrives at the Skid Strip from NAR on August 12.

The CM's environmental control unit precedes the spacecraft onto the lift.

A crate of support equipment is also transferred onto the lift. The message on the CM cover is intended for the McDivitt crew, still slated to fly the mission. The "I know you can fly a spacecraft better than can you play baseball" banner is a reference to a competition between NAR and astronauts teams played on Merritt Island.

NAA workers in a Hi Ranger bucket crane prepare to unpack the CM in the MSOB's high bay on August 12.

NAA workers remove protective shipping panels from the CM.

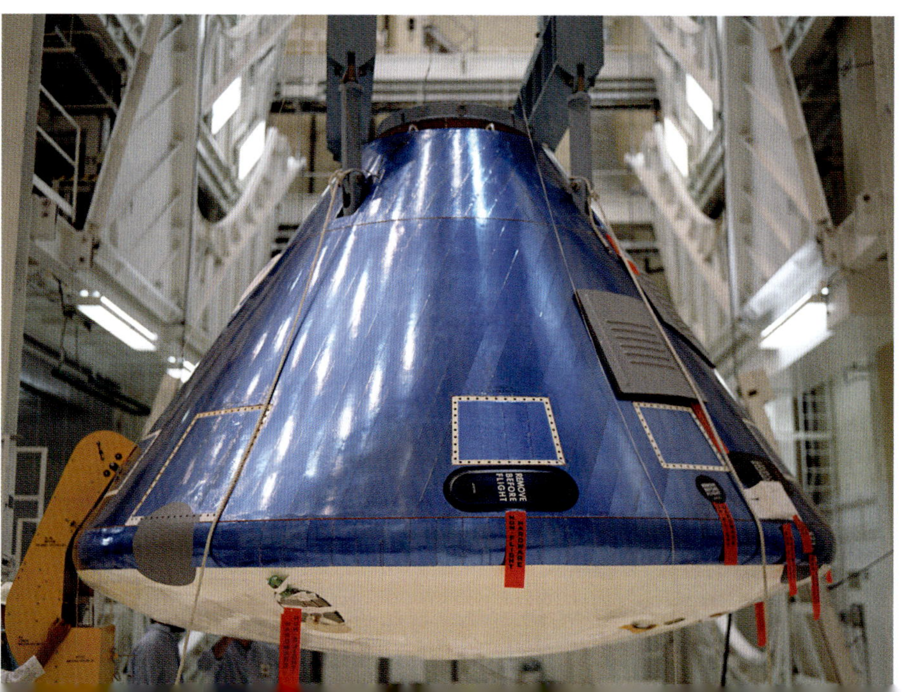

A bridge crane moves the CM to a temporary workstand on August 12 for checkout until it is mated with the SM in altitude chamber L.

On August 14, McDonnell Douglas technicians inspect the S-IVB's aft interstage attach ring in the VAB's transfer aisle.

A bridge crane moves the S-IVB toward the second stage, stacked on ML-1 in high bay 1.

The S-IVB with its J-2 engine is moved toward the high bay.

NAA technicians inside the top of the S-II prepare for mating the S-IVB to the S-II stage.

Technicians guide the skirt down by hand.

A bridge crane is attached to the IU's handling fixture in the transfer aisle, parked on the transfer pad.

Soon after, the crane lifts the instrument unit (IU) from its dolly in the transfer aisle. The Apollo Facilities Vehicle, first used with the SA-500F "dummy" Saturn V, is at right.

IBM technicians inside the S-IVB's forward skirt watch as the IU is lowered. Service arm no. 7 is at right.

Bendix and IBM managers watch as the IU is inches from mating. The technicians inside will exit through the open hatch at left.

The Apollo Facilities Verification Vehicle is moved from the transfer aisle toward high bay 1 for stacking on August 16. It will be temporarily used for tests involving the LUT's service arms.

The CM/SM/SLA is lowered for mating to AS-503. Three days later, facing LM delays, NASA managers decide to cancel the booster's planned mission, reassigning it to launch CSM-103 and the LTA-B ballast. No decision is made about the specifics of the new mission, set for December.

An MSOB bridge crane is lowered toward CM-103 to lift it from its workstand into altitude chamber L on August 21.

CM-103 is lowered into the chamber for mating with SM-103.

Fuel tanks line the inside of the RCS quad B SM panel as it's lowered in the chamber August 22, with the CM at right.

Three of the four RCS nozzles are visible as the panel is installed.

Paint on the CM's Launch Escape System (LES) tower peels off when the protective plastic wrap is removed on August 23 in the Pyrotechnic Installation Building at KSC. The welded titanium structure will be repainted.

CSM-098 is inside chamber A at MSC's Space Environment Simulation Laboratory (SESL) before a September test to certify a CSM for lunar spaceflight. The modules had been tested in June as 2TV-1 to qualify the spacecraft for Earth orbit (Apollo 7). This 120-foot-high chamber can simulate pressures and temperatures of up to 130 miles in altitude.

Chamber technicians help a crewmen ingress on September 4. He wears a training suit for Joe Engle, who'd participated in the June test. A critical modification from June was adding the new unified CM hatch. Tool B, a hand crank to operate the hatch from the outside, is in place.

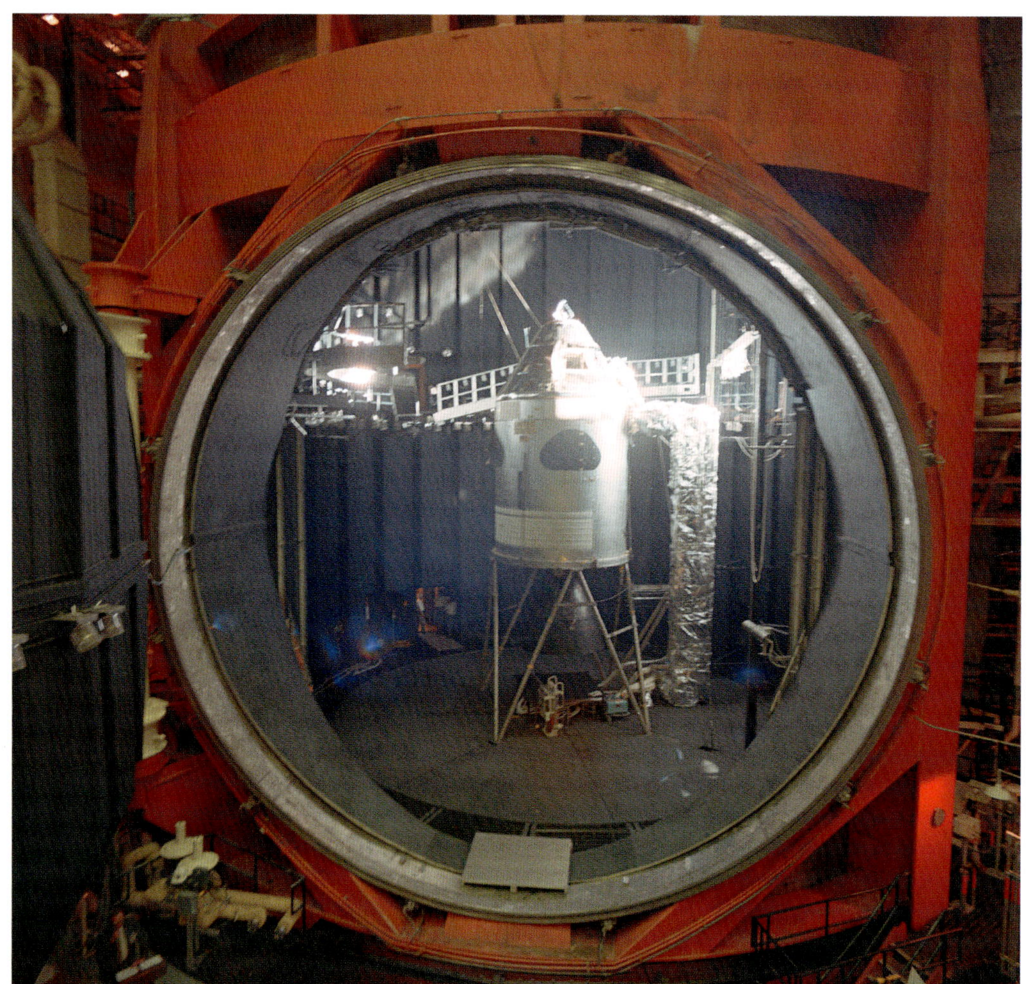

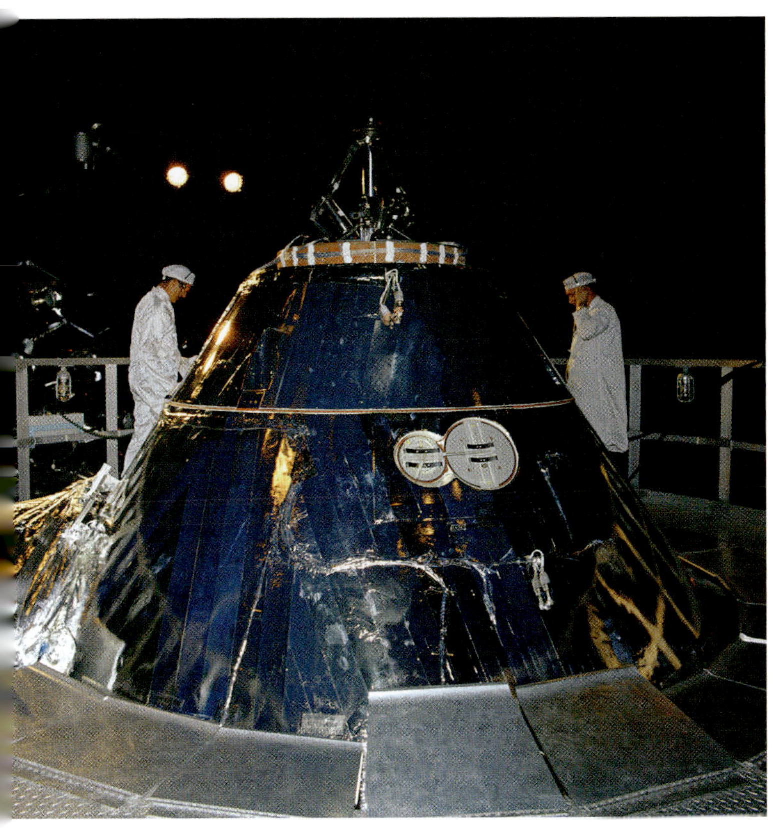

The techs make final checks before clearing the SESL. Covers protect the CM's sextant and telescope openings. Modifications include adding an LM docking probe (*top*), which would first be used on Apollo 9. Carbon-arc lamps simulate solar heating, and liquid nitrogen in the chamber's walls simulates the cold of cislunar space.

Crew members get settled inside the CM before the hatch is closed, with the middle couch removed to provide more room during the test. The crew successfully performs a simulated EVA on the second day by depressurizing the cabin, opening the hatch, and repressurizing the cabin.

Left to right: Lloyd Reeder, Turnage Lindsey, and Alfred Davidson, all USAF majors assigned to MSC's Flight Crew Support Division, wear parkas and Beta cloth outfits on September 9 after spending five days in the chamber.

The complete 33-foot-long LES, manufactured by Lockheed Propulsion Co. of Redlands, California, undergoes processing in the Pyrotechnic Installation Building at KSC on September 6. Lockheed provides both the launch escape motor and the pitch control motor. One of the two canards will cover the opening at extreme right.

Ordnance technicians will install the igniter assemblies for the three LES motors (jettison, escape, and pitch control) through access panels.

One of the two oval solid-fuel tower jettison motors is sealed with thermal insulation in the foreground. The motors are built by the Thiokol Chemical Corp. of Elkton, Maryland. The CM's apex cover is in the background.

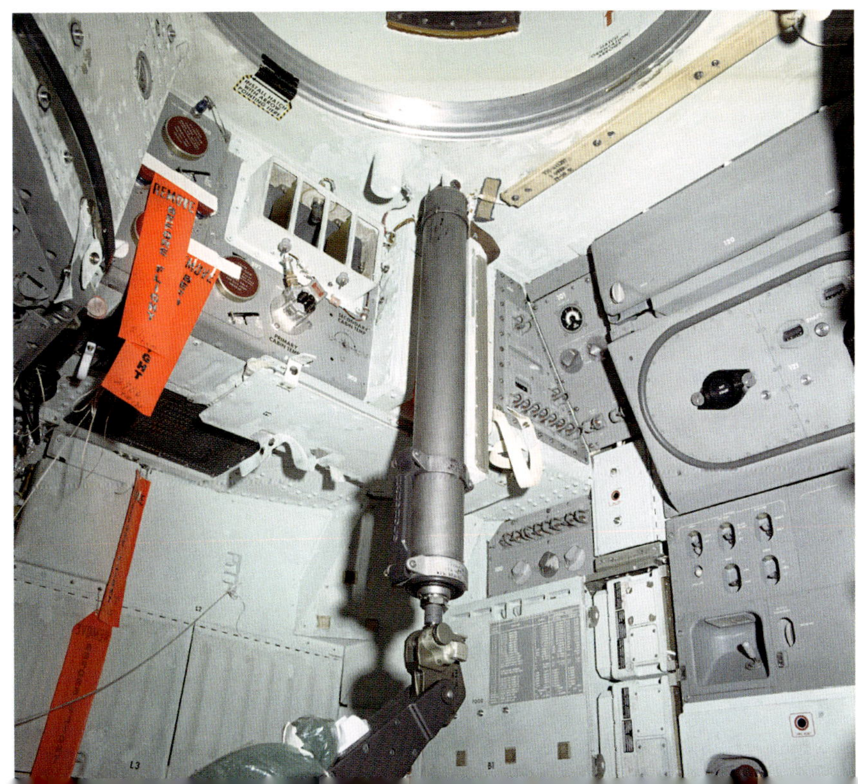

The CM's open hatch shows its Block II latch mechanism in altitude chamber L in the MSOB on September 7.

The LM pilot's side of the CM cabin in the chamber on September 7. A TV camera mounted in a side window will provide video for the test.

This photo from the lower equipment bay looks toward the docking tunnel at top.

On September 11, Borman gets suited for a simulated (sea-level) chamber test of the CM.

Lovell's suit is pressurized for the chamber run.

Anders prepares for the test.

Anders (*foreground*) and Lovell wait to enter the CM.

NAR technicians help Borman ingress.

Borman (*left*), in the commander's couch, is strapped in, with Lovell at right.

Lovell following the test

Anders and Lovell after the test, which is repeated by the backup crew the next day. The tests will be conducted at altitude by the prime crew on September 20 and the backup crew on September 22.

The CM is prepared for tests at altitude in chamber L on September 13.

An RCS quad just below the umbilical structure on the SM is ready for installation in the chamber on September 13.

This ground support panel regulates gaseous nitrogen and helium used to purge and pressurize the SPS fuel lines during chamber tests.

The CSM is moved from the altitude chamber to a test stand in the MSOB on September 24. The missing panels at left reveal two of the three fuel cells, with their oxygen and hydrogen supply tanks below.

A replacement SPS engine is ready for installation on September 25. The new engine is qualified for the low temperatures of lunar flight, necessary for a potential Moon mission.

One of three SM fuel cellfuel cell powerplant assemblies, manufactured by the Pratt & Whitney Aircraft Co. of East Hartford, Connecticut, on September 27 before installation. They combine liquid hydrogen and liquid oxygen (LOX) to generate electrical power and produce potable water as a byproduct.

One of three tightly packed main parachutes, part of the CM's Earth Landing System, is photographed during CM closeout on September 27.

The docking ring, which is not needed without a LM, is removed from the CM on September 27.

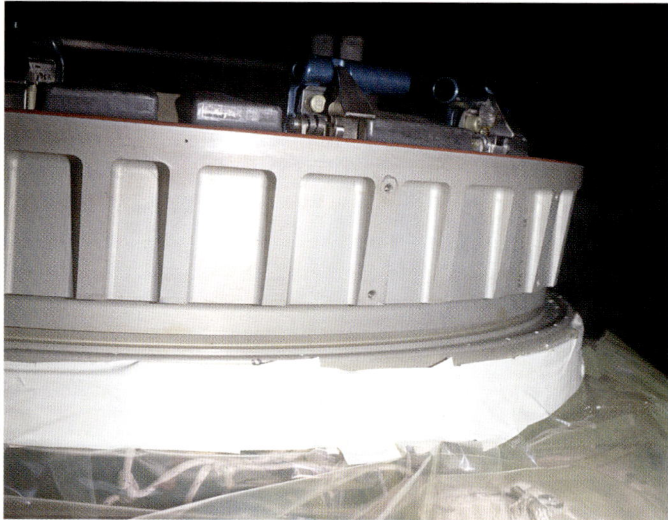

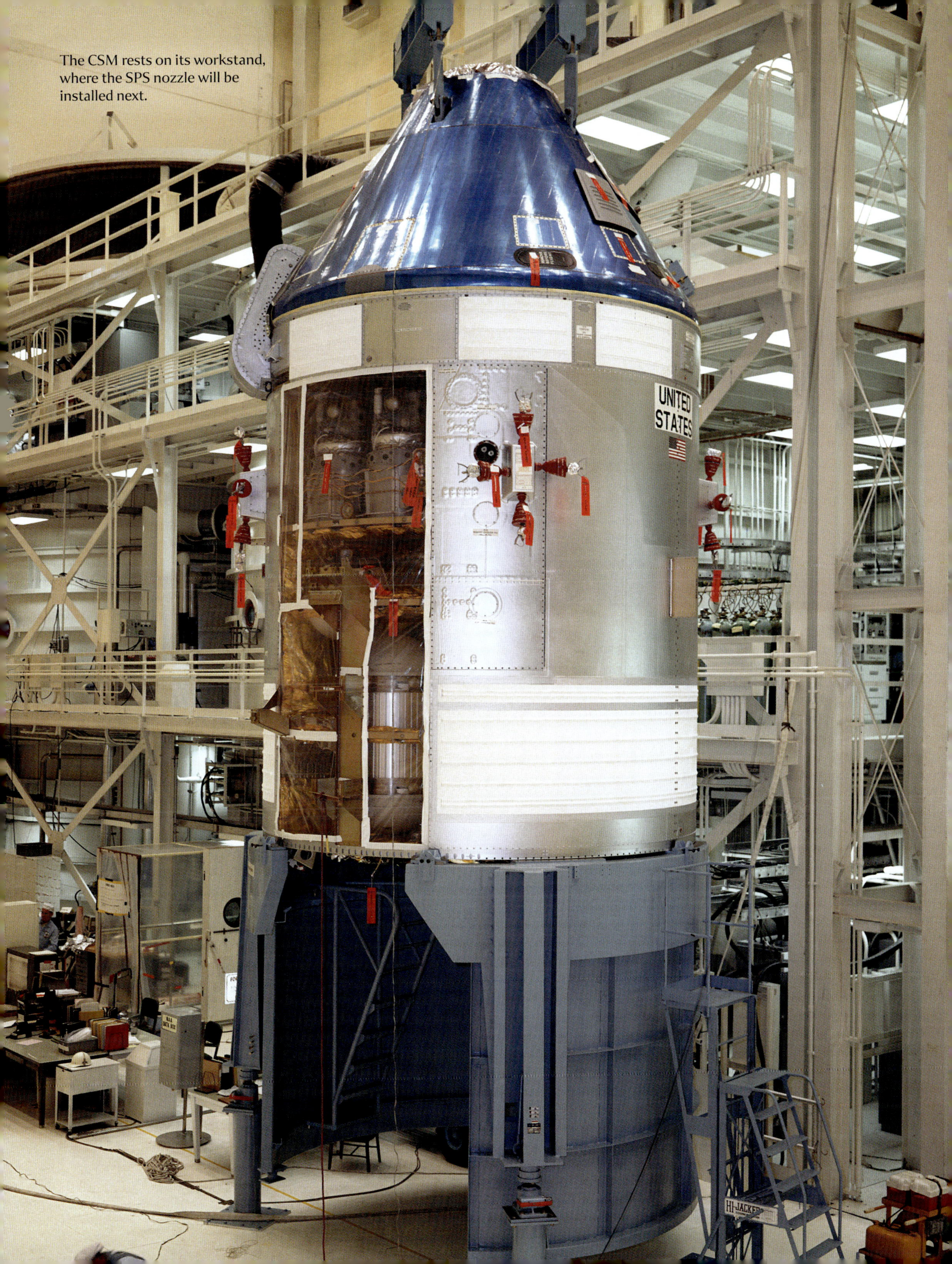

The CSM rests on its workstand, where the SPS nozzle will be installed next.

CHAPTER 6

October 1968

On October 2, work platforms and test equipment surround the forward end of Lunar Module Test Article B (LTA-B), mounted in its SLA in integrated workstand no. 1 in the MSOB. The structure serves as an LM substitute but, at 19,900 pounds, is 60 percent lighter than a flight LM.

A view looking up at LTA-B from the aft end of the SLA. It is instrumented to measure acceleration forces.

LTA-B is mounted to the same four attach fixtures the LM will use, beginning on the next mission.

On October 4, a bridge crane in the MSOB's high bay lifts Apollo 8's CSM from workstand no. 4 for mating with SLA-II.

Festooned with red "Remove Before Flight" streamers, the CSM's move is monitored by NAR technicians.

A high-gain antenna, which would be used for unified S-band communications at lunar distances, is installed on the SM to preserve a Moon mission option. It consists of an array of four 31-inch parabolic dishes clustered around a square wide-beam feed horn.

The CSM is lowered toward its SLA in the workstand.

Early on the morning of October 7, work platforms are raised after the spacecraft stack has been wrapped for transport to the VAB.

The CSM and SLA are at the MSOB door, ready to be towed, as NAR and Bendix technicians watch in the foreground.

The CSM/SLA is escorted along the Saturn Causeway to the VAB at 7:00 a.m. EDT. The roof of the LC-39 Press Site grandstand is visible at right.

The stack passes a VAB parking area.

Flanked by KSC Security cruisers, the spacecraft arrives outside the VAB's south door at 7:20 a.m.

Workers using the white bucket crane at left attach a bridge crane to the CM.

A bridge crane moves the CSM/SLA past high bay 2 at left, where stacking for Apollo 10 will begin in December.

Managers on the Level 6 work platform (*upper center*) watch as IBM technicians in the IU prepare to mate the spacecraft.

The CSM/SLA is carefully lowered for mating. The collapsed white room is at the end of service arm no. 9, to the right of the CM. Three technicians work at an S-IVB access panel below.

The same day at Pad A, technicians test the water deluge system for the flame trench. The slide at left, leading to the emergency escape bunker, is normally connected to a chute in the ML. The MSS is in place for this test at right.

Water floods the flame trench during the test.

The flame trench walls are lined with Fondu Fyre™, a heat- and erosion-resistant concrete developed for this use. The excess water flows into a "burn pond" in the background. A flame deflector can be seen at center through the mist.

On the morning of October 9, Apollo 8 is ready for rollout to Pad A from the VAB as Charles Mathews, deputy associate administrator for manned space flight (*left*), chats with Robert Gilruth, MSC director.

Lovell (*right*) talks with RAdm. Roderick Middleton (USN, ret.), KSC Apollo manager (*left*), and Eberhard Rees, MSFC deputy director, as they wait for the rollout.

The crew poses with key NASA executives as the slow rollout begins about 8:30 a.m. *Front row, left to right*: Lovell, Anders, and Borman. *Standing, left to right*: Gilruth; George Mueller, NASA associate administrator for space flight; Rees; and Kurt Debus, KSC director.

Anders, Borman, and Lovell stand between tracks on the crawlerway as the move gets underway one day ahead of schedule.

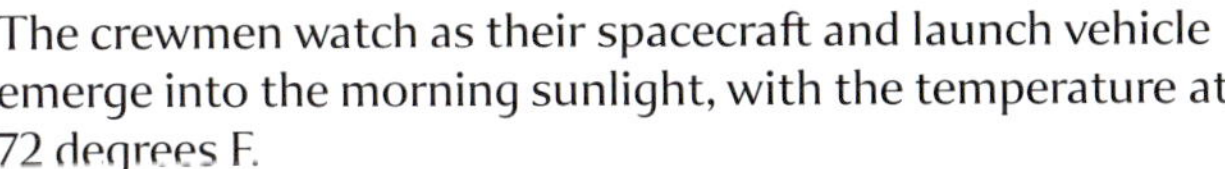

The crewmen watch as their spacecraft and launch vehicle emerge into the morning sunlight, with the temperature at 72 degrees F.

Borman (*left*) speaks to reporters and photographers with Lovell and Anders.

Lovell, Anders, and Borman pose for photographers.

The astronauts depart after the photo opportunity.

Debus comments during the rollout.

Mueller; Chris Kraft, flight operations director at MSC; Charles Donlan, deputy associate administrator for manned spaceflight; and Mathews pose in front of Apollo 8.

Members of the news media photograph the scene. *At right*, ABC News science editor Jules Bergman visits with Borman's wife, Susan.

A top-down look from the VAB roof

This view from the VAB shows the stack heading toward Pad A at less than 1 mph, with the MSS in the distance.

The Apollo 8 Saturn V gradually moves away from the VAB on the crawler-transporter. The combined spacecraft, launch vehicle, and ML weigh 12 million pounds.

The stack creeps past the LC-39 turning basin, where its first two stages had arrived by barge ten months earlier.

The ML's hammerhead crane (*top*), four of its nine service arms, and a number of technicians on the LUT can be seen in this photo from the parked MSS.

Apollo 8, with the MSS at left

Apollo 8 passes the crawlerway fork to Pad B (*left*), with Pad A and the Atlantic Ocean ahead in the distance.

Service trucks and a fire engine on the Saturn Parkway, parallel to the crawlerway, accompany the ML.

The stack passes pine trees on its slow journey.

The crawler-transporter inches its way up the ramp, using a laser guidance and hydraulic leveling system to keep the tip of the CM within about 1 foot of level as it moves up the 5 percent grade.

The crawler moves over a below-grade sensor that will track two prisms in the IU to ensure the Saturn is on its precise trajectory at launch. A third prism mounted on the IU accounts for vehicle sway. The pad's LOX storage tank is at right.

Apollo 8 nears the hardstand in this view looking east.

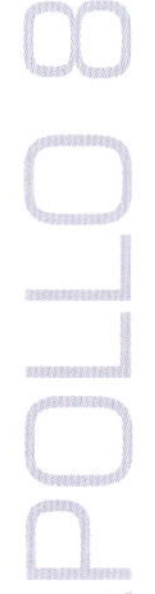

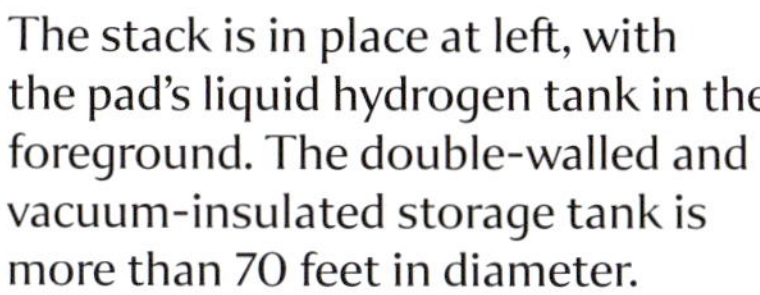

The stack is in place at left, with the pad's liquid hydrogen tank in the foreground. The double-walled and vacuum-insulated storage tank is more than 70 feet in diameter.

The Apollo 8 Saturn V is locked down on the hardstand pedestals that afternoon after a six-hour trip.

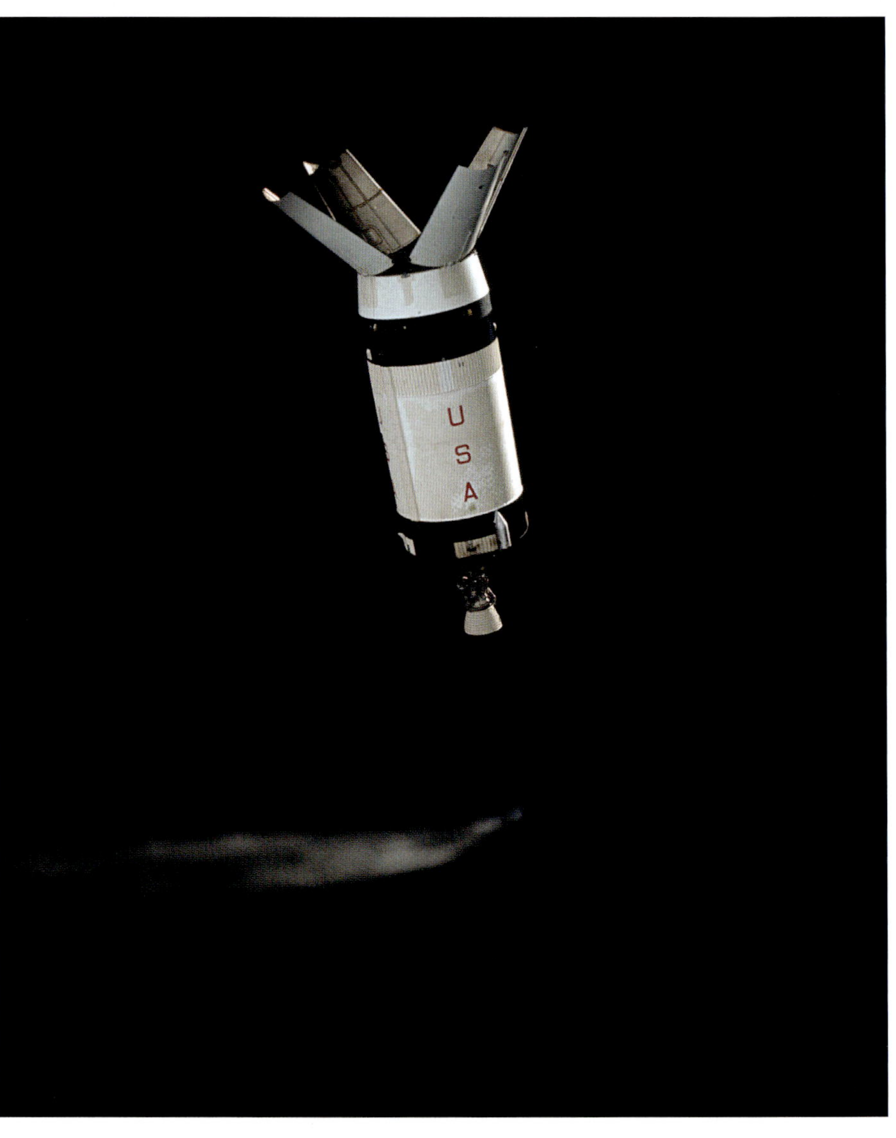

On October 11, Borman and his wife, Susan (*left*), wait for the Apollo 7 launch at the VIP viewing area on the East NASA Causeway along the Banana River. They're joined by Apollo 8 support crewman Jerry Carr (*next to Susan*), Al Worden (*holding drink*), and Carr's wife, JoAnn.

Apollo 7 launches into orbit on a Saturn IB at 11:03 a.m. EDT from LC-34 at Cape Canaveral AFS, on a nearly eleven-day shakedown flight for the CSM.

The Apollo 7 astronauts get as close as 70 feet to their S-IVB second stage after a second rendezvous on October 12. The CSM stays there for half an hour.

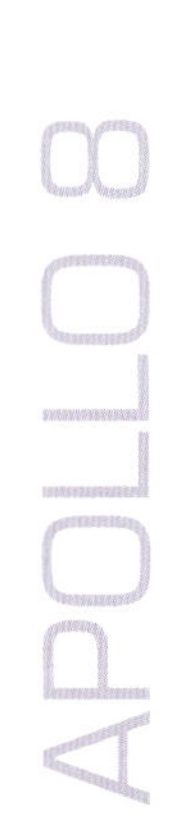

The Apollo 7 CM is ready to be hoisted aboard USS *Essex* about two hours after splashdown, 300 miles south of Bermuda in the Atlantic Ocean on October 22.

Anders (*center*) examines a Hyatt model of the CM with Lovell (*left*) and Borman before an October 18 meeting in MSC's Building 4. Photos of the Apollo 1 astronauts are on the wall behind them.

Prime crew commander Borman and CM pilot Lovell (*left*) meet with their counterparts on the backup crew (Armstrong and Aldrin) and support crew (Brand and Carr) to review planned SPS firings during the mission. Two paintings above the bookcase depict the Apollo I astronauts.

Original Apollo 8 CM pilot Mike Collins also attends the meeting.

Borman during the meeting, which also covers mission rules

Lovell makes a point.

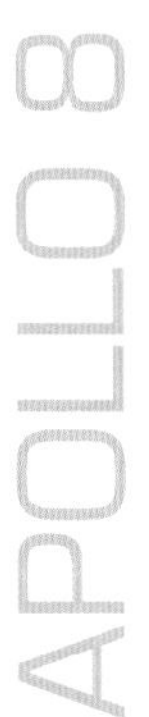

On the afternoon of October 22, the crewmen are met by NASA engineering manager Jim Ragusa as they exit the Astronaut Transfer Van for slide-wire training at Pad B, using a temporary tower built just for this use.

Borman, Lovell, and Anders talk with NASA and Bendix managers before the training. They'll stand on the wooden stair to try out a harness before jumping from the 35-foot tower at right.

Lovell and Anders. Their Saturn V can be seen on Pad A in the distance.

Borman tells reporters he'd like to try the slide-wire again with his helmet on, to make sure it wouldn't interfere with the harness.

Lovell during the training

Anders adjusts his glove.

Borman dangles from the harness as Ragusa (*jumpsuit*) supervises.

Lovell takes his turn during the training.

Anders tries out the harness.

Top to bottom: Borman, Lovell, and Anders climb the tower for slide-wire runs.

Borman has slide-wire harness attached as Lovell (*right*) waits.

Lovell comes down the wire as Ragusa reaches up.

Left: Borman gets ready for a photo opportunity along Fluid Servicing Road, about 2 miles west of Pad A on the way back to the MSOB.

Two below: Borman, Lovell, and Anders pose for photos. KSC public information chief Jack King (*third from left*) squats.

Borman, Lovell, and Anders pose with their Saturn V in the distance.

Later that afternoon, backup crewmen (*left to right*) Haise, Armstrong, and Aldrin participate in the same slide-wire training.

Aldrin briefly dons his helmet to check for any issues.

Aldrin (*partially hidden*), Haise (*back to camera*), and Armstrong before their slide-wire runs

Aldrin slides down as Ragusa and Armstrong look on.

Haise (*right*) after his slide-wire run. He'd just moved from the support to the backup crew and wears Borman's training suit since he didn't have his own yet.

Left to right: Aldrin, Haise, and Armstrong head back to the transfer van to return to the MSOB.

The following day, October 23, suit tech Joe Schmitt (*left*) works with Borman in the MSOB before an emergency egress drill at the pad.

Anders adjusts his microphone.

A suit tech helps Lovell with a wrist ring.

Anders is ready to don his helmet.

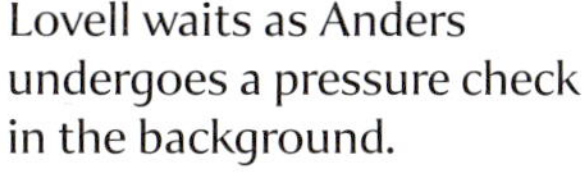

Lovell waits as Anders undergoes a pressure check in the background.

Borman begins to prebreathe pure oxygen remove nitrogen from his bloodstream to prevent the bends, as do Lovell and Anders.

Lovell, Borman, and Anders depart for the trip to Pad A.

The astronauts head toward the ML elevator at the pad, with blue protective covers on their helmets.

A rescue team in the white room closes the CM hatch before the simulated emergency.

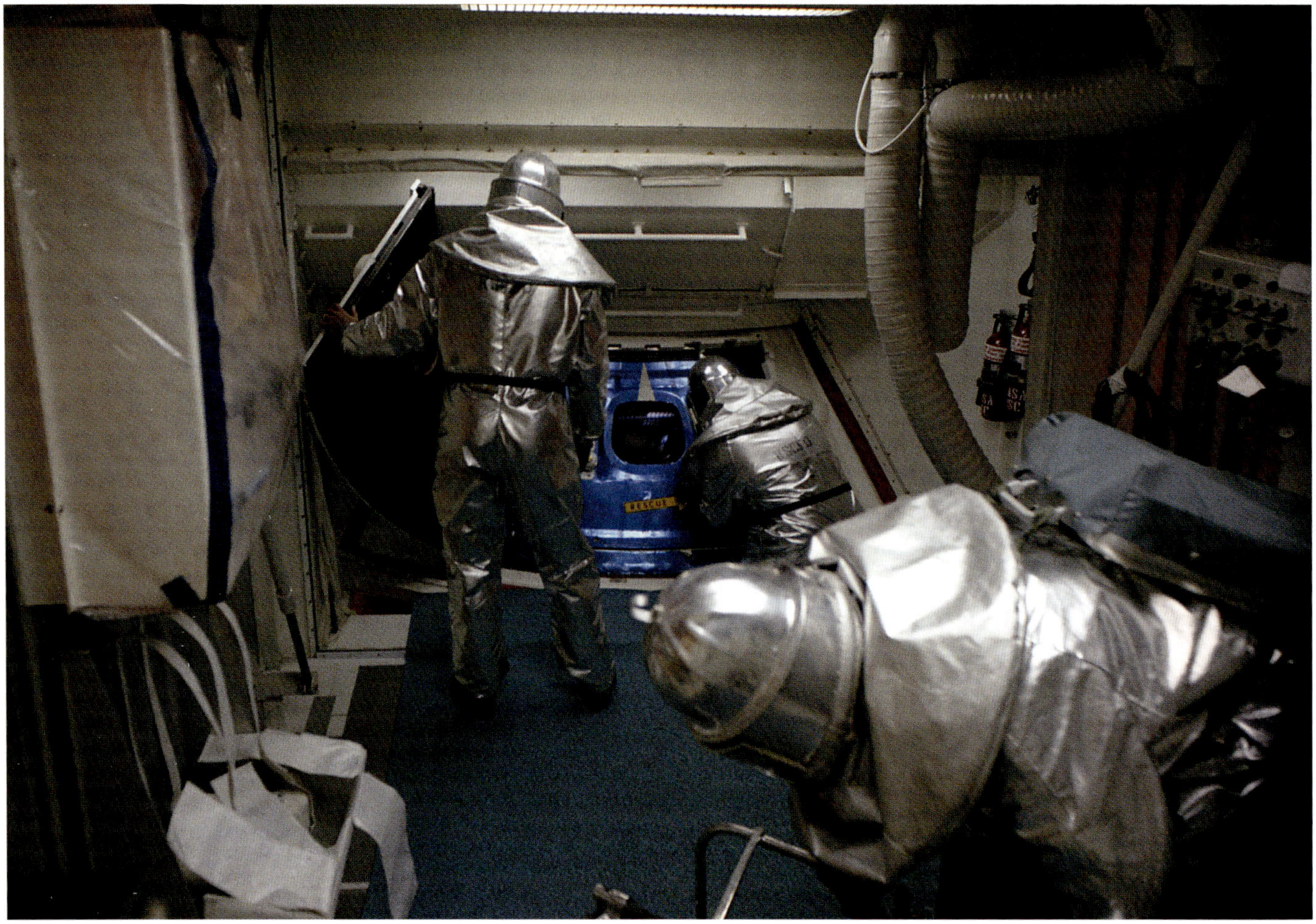

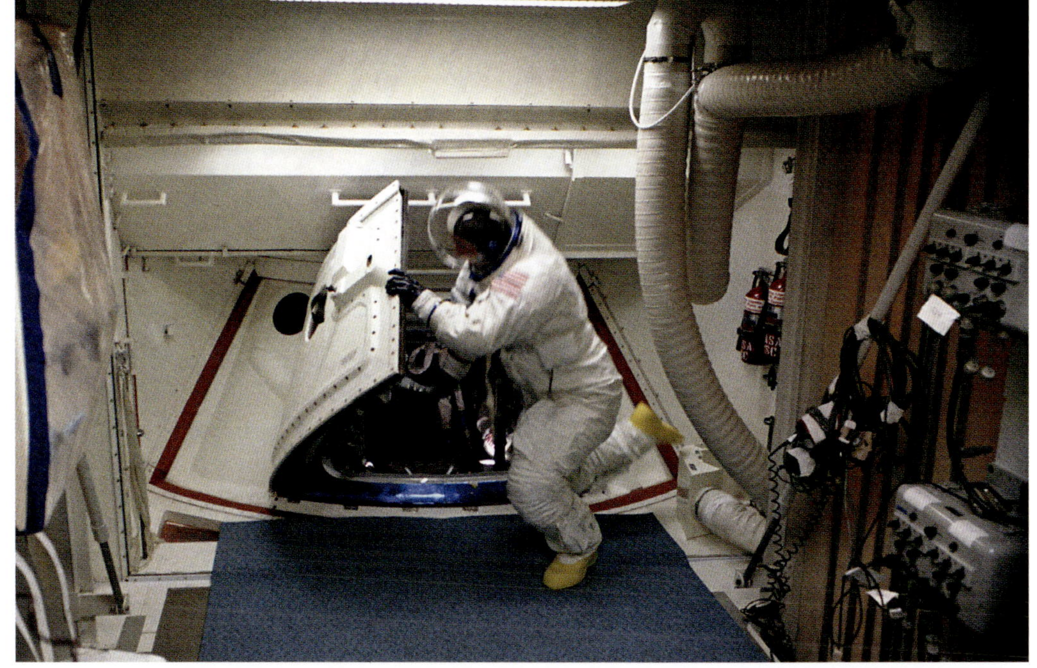

Lovell holds the boost protective cover hatch as he climbs out of the CM first.

Lovell dons an emergency oxygen pack.

Borman reaches for an oxygen pack, with Anders behind him.

Borman (*center*), Lovell (*foreground*), and Anders evaluate the simulated emergency exercise with a stand-in pad leader (*left*).

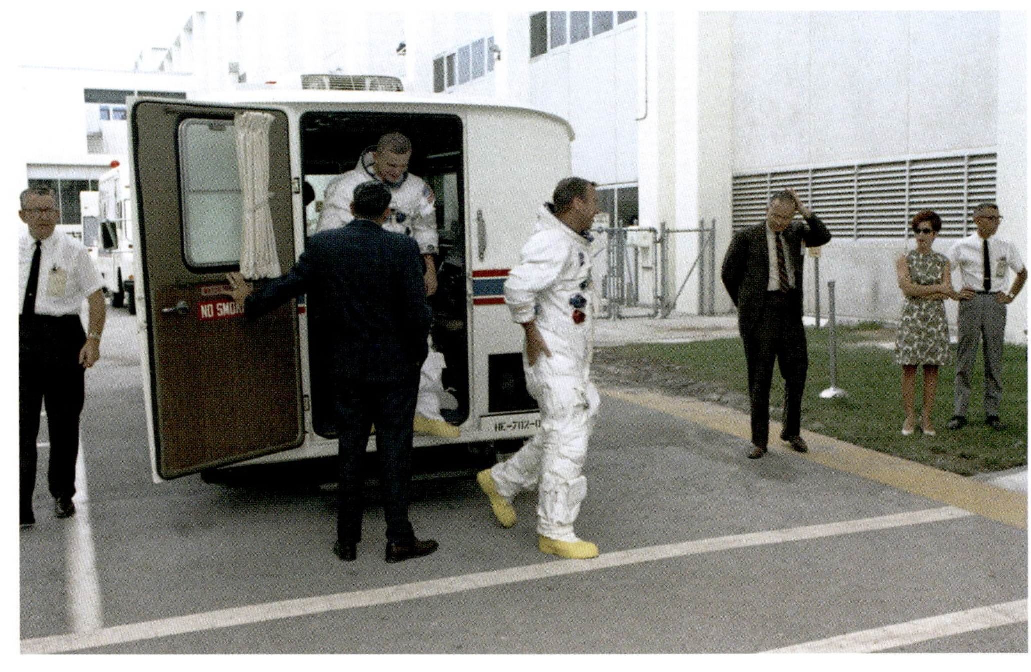

Borman (*left*) and Lovell return to the MSOB after the egress drill; security officer DeArmond Matthews holds the door, with Anders still in the van. Driver Steve Tatham of KSC Security walks back at far left.

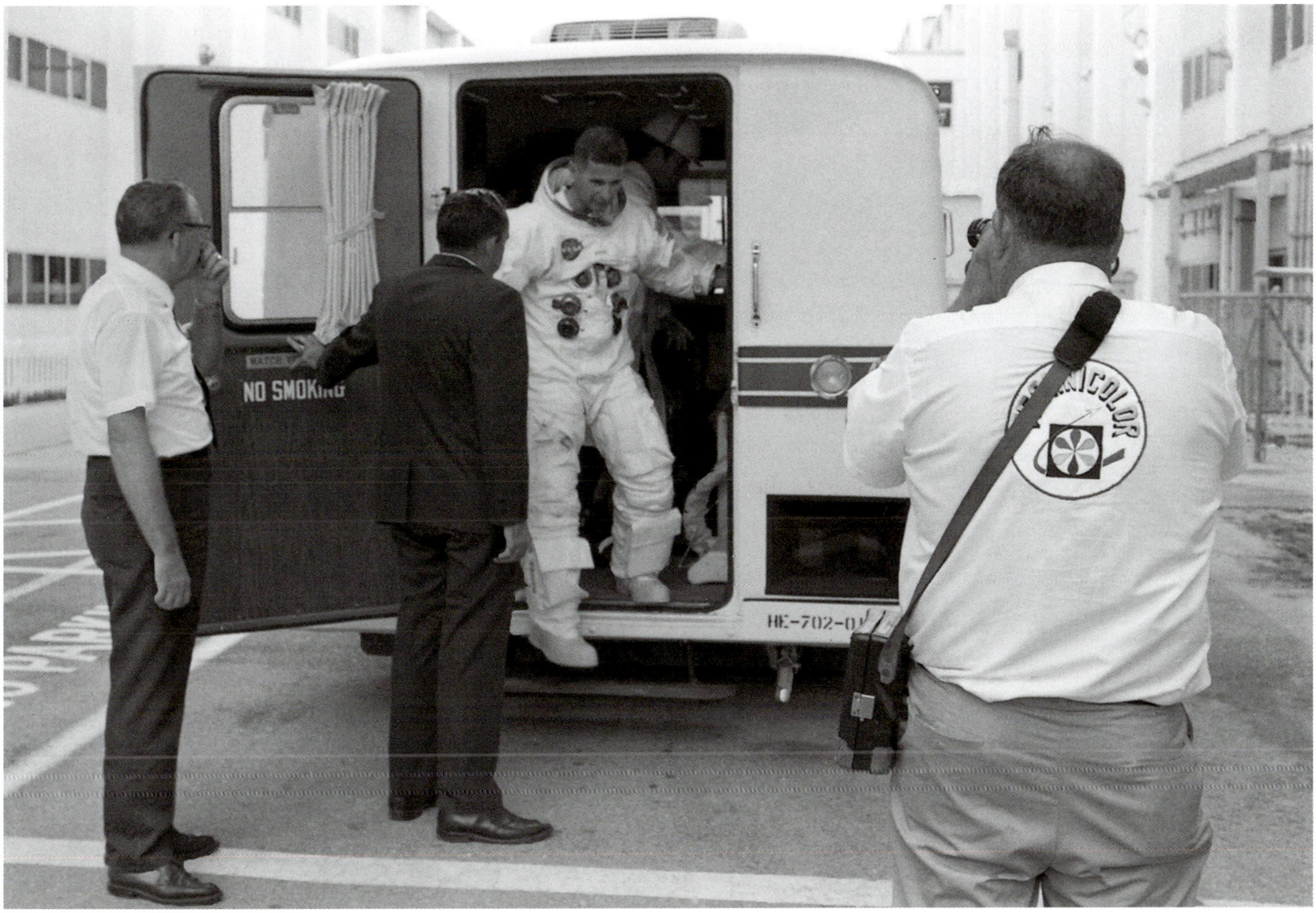

Anders exits the van as NASA contract photographer George Nevin captures the scene.

Later that day, the backup crew, led by Armstrong, will repeat the drill.

Haise (*left*) works with suit tech Dave Halfhill.

Aldrin is suited for the emergency egress test, with Haise in the background.

Armstrong is assisted by suit tech John Bolan.

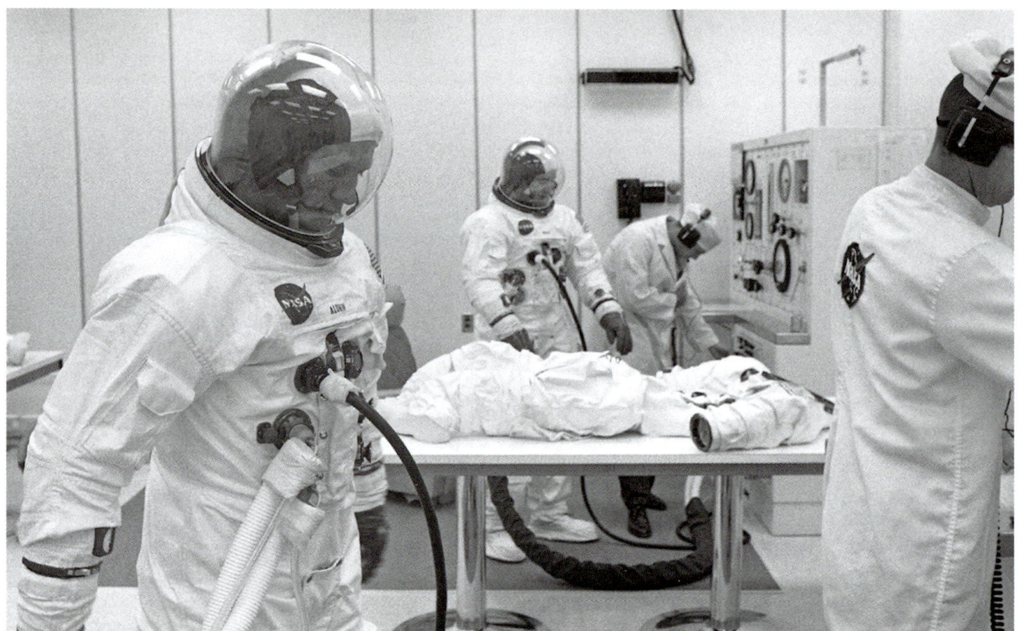

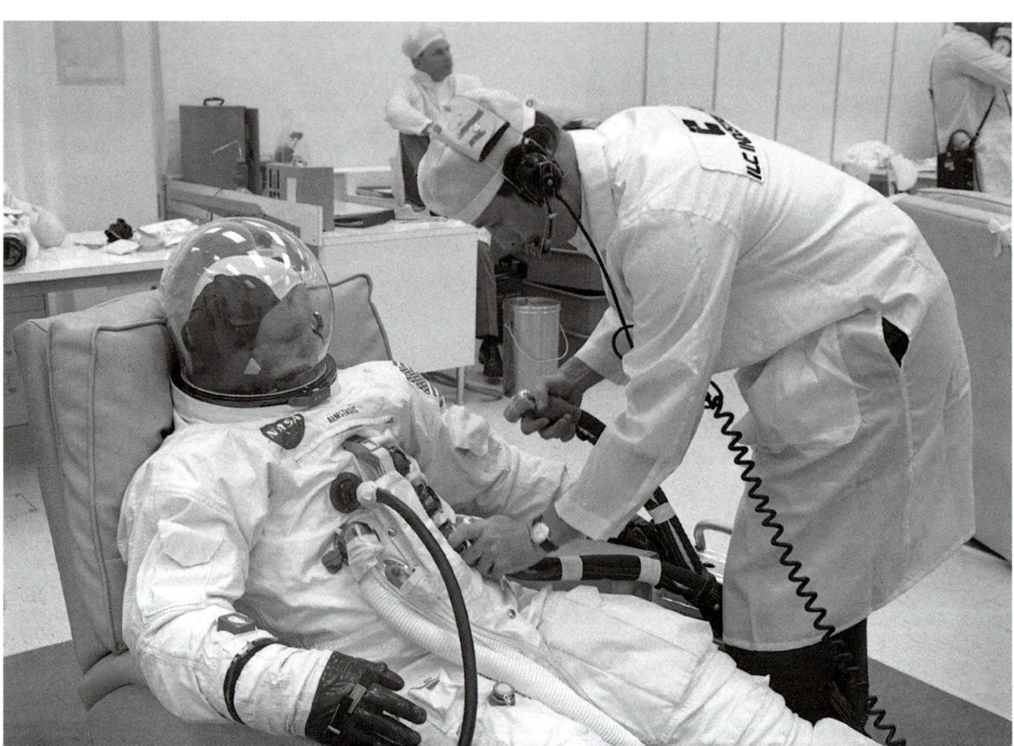

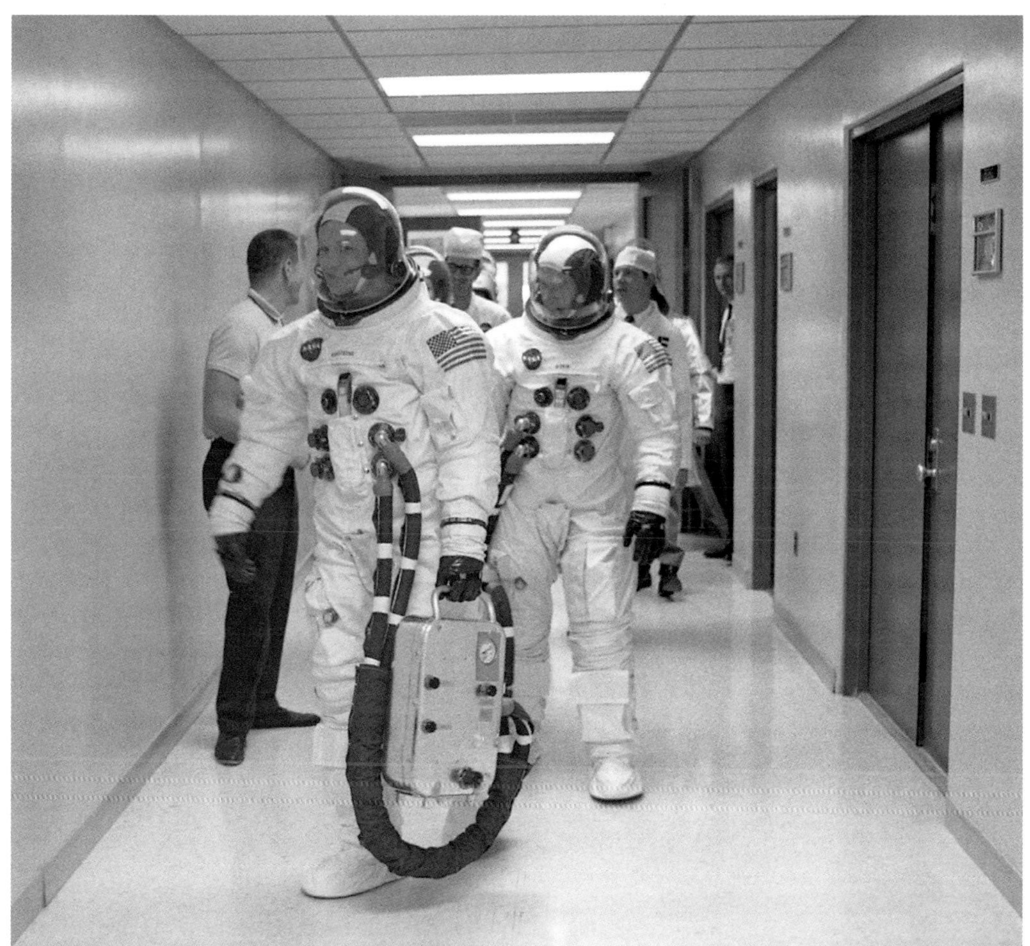

Armstrong, Aldrin, and Haise (*partially visible*) depart the suit room at the MSOB. Astronaut Jack Swigert (*left*) is on hand to welcome the Apollo 7 crew back after their mission.

Armstrong leads Aldrin and Haise to the transfer van to Pad A.

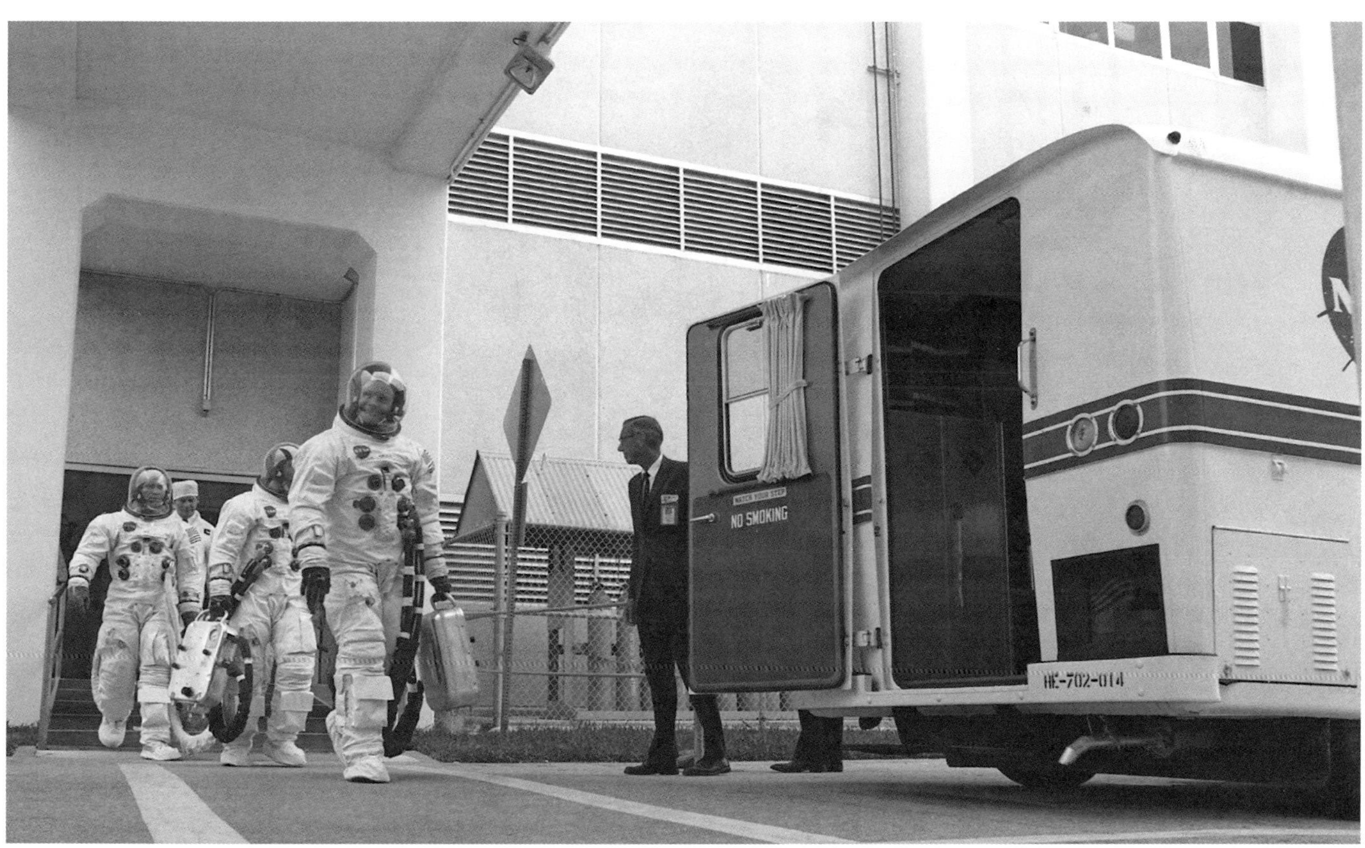

On October 25, Anders, Borman, and Lovell prepare for recovery training aboard NASA's Motor Vessel (MV) *Retriever* in the Gulf of Mexico, 8 miles off Galveston, Texas. They had met with the Apollo 7 crew at KSC the day before.

Lovell aboard *Retriever*

Borman, Lovell, and Anders pose with CM boilerplate (BP)-1102A aboard *Retriever*. BP-1102A is an aluminum boilerplate built to test the CM's flotation characteristics.

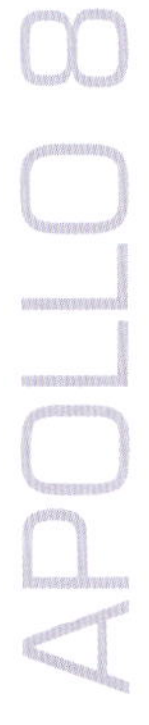

Borman is suited below deck aboard *Retriever*. The pressure suits do not have the white Beta cloth outer cover.

Anders aboard *Retriever*

Borman and Anders are ready to begin.

ILC suit tech Bob McDilda (*left*) helps Lovell ingress the CM boilerplate.

Lovell's wife, Marilyn, watches the training from *Dutchess*, a yacht owned by La Porte, Texas, businessman Paul Barkley and loaned as an observation boat for journalists covering the exercise. "Wouldn't it be nice if they were just coming back from the Moon right now, and it was all over?" she asks a reporter. *Photo by Fred Kaufman / AP*

Left to right: Anders, Lovell, and Borman aboard the boilerplate

A US Coast Guard Sikorsky HH-52A Seaguard helicopter lifts one astronaut in a Billy Pugh net. The crewmen have inflated the boilerplate's uprighting bags after it was dropped apex first into the Gulf of Mexico.

CHAPTER 7

November 1968

Anders, Lovell, and Borman pose in front of the centrifuge gondola in the Flight Acceleration Facility in Building 29 at MSC on November 1. The gondola is on the end of a 50-foot arm, which swings the gondola to duplicate the g-forces the crew will experience during launch and reentry.

Commander Frank Borman is suited for a session in the centrifuge. It was built primarily for Apollo astronaut training.

LM pilot Bill Anders and CM pilot Jim Lovell chat before entering the centrifuge's gondola.

Borman, Lovell, and Anders aboard the gondola

Anders climbs from the gondola after the training run.

Borman is last out, helped by tech Bob McDilda from International Latex Corp. (ILC) of Dover, Delaware, the pressure suit contractor.

On November 4, recently retired NASA administrator Jim Webb (*sixth from left*), launch director Rocco Petrone (*right of Webb*), and KSC director Kurt Debus (*far right*) tour Pad A with a visiting group of American and British scientists. Two of the ML's three tail service masts are at right, with the third partly visible at far left.

The Apollo 8 prime and backup crews meet in the astronaut quarters in the MSOB on November 5. *Clockwise from bottom*: Vance Brand (support crewman), Lovell, Haise, Anders, Borman, Aldrin, Armstrong, and two training managers.

The map behind Aldrin (*left*) and Armstrong indicates photography targets for Apollo 7.

Lovell, Haise, Anders, and Borman

Clockwise from left: Brand, Lovell, Haise, Anders, Borman, Aldrin, and Armstrong. Astronaut Jack Schmitt (*foreground*) is a geologist who provides input on lunar observations.

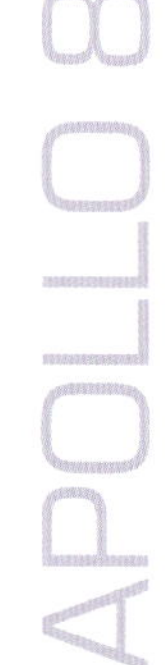

The decision to send Apollo 8 to the Moon was officially announced on November 12 but had its genesis months earlier. The idea had first been proposed by Apollo Spacecraft Program Office manager George Low at MSC when it became apparent the LM for the mission would not be ready in time. On August 9, Low outlined his idea to center directors from MSC, KSC, and MSFC at a meeting in Huntsville. Others attending included flight director Chris Kraft, Astronaut Office chief Deke Slayton, launch director Rocco Petrone and Apollo program director Sam Phillips. Everyone agreed that while it would be challenging, they couldn't identify any real showstoppers. NAR's Space Division president, Bill Bergen, however, was not receptive when contacted.

Low and the center directors met with officials at NASA Headquarters the following week, and administrator Jim Webb, although initially negative, soon approved the plan, assuming Apollo 7 was a success in October. On the morning of August 19, acting administrator Tom Paine and Phillips announced at a headquarters news conference that the LM would no longer be part of the Apollo 8 flight and that the agency was considering various mission objectives.

On September 14, citing informed sources, UPI reported from Houston that the mission would orbit the Moon ten times at Christmas. On November 12, Paine and Phillips made the official announcement at a headquarters news conference that the Apollo 8 astronauts would leave Earth orbit for the first time and head for the Moon.

Low

Phillips

Paine

On the morning of November 12, mission director William Schneider (*left*) and Apollo program director Sam Phillips announce at NASA Headquarters during a three-hour news conference that Apollo 8 would orbit the Moon. "We're ready for a lunar orbital mission, and we expect to carry out the entire mission. If I was not convinced, I wouldn't have recommended it in the first place," Phillips says. *AP photo*

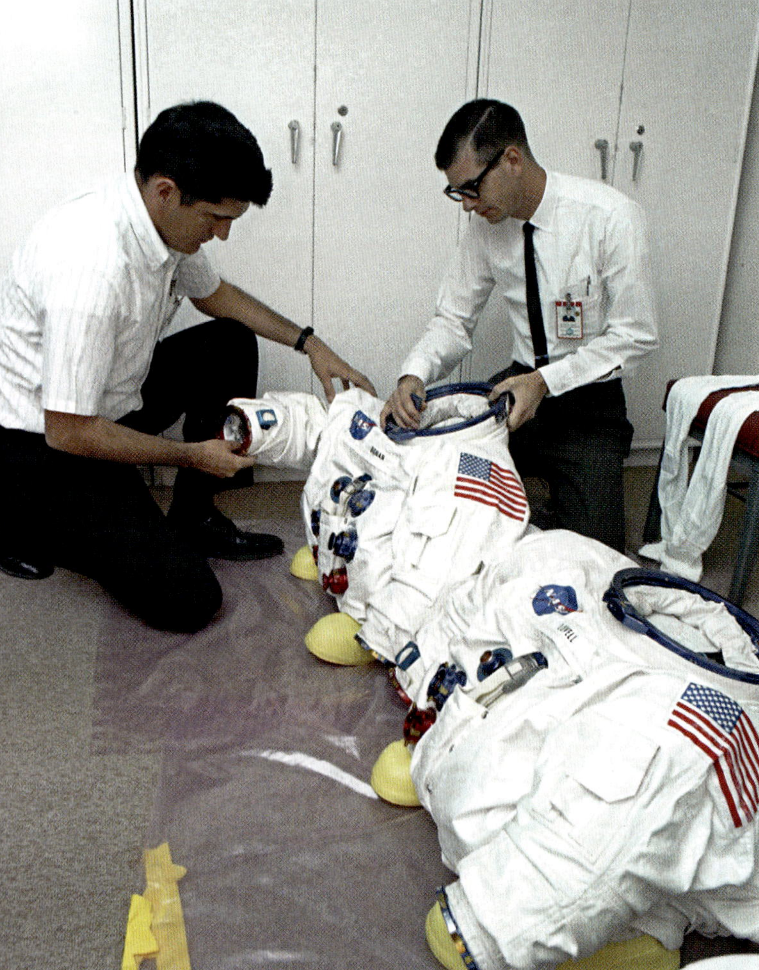

On November 13, ILC suit tech Bob McDilda (*left*) and NASA's Dick Sandridge examine Borman's pressure suit, with Lovell's at right, in the Flight Crew Training Building, for crew photos and simulator training. The astronauts wear ILC's A7L suits.

Sandridge (*left*) and McDilda prepare Anders's suit.

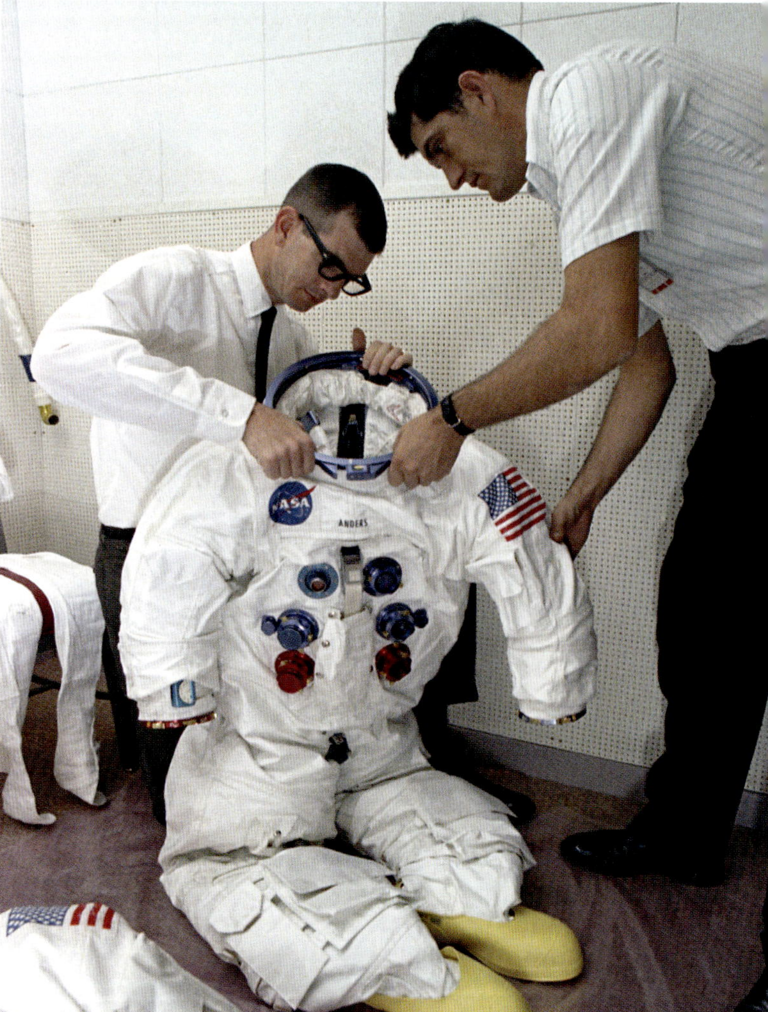

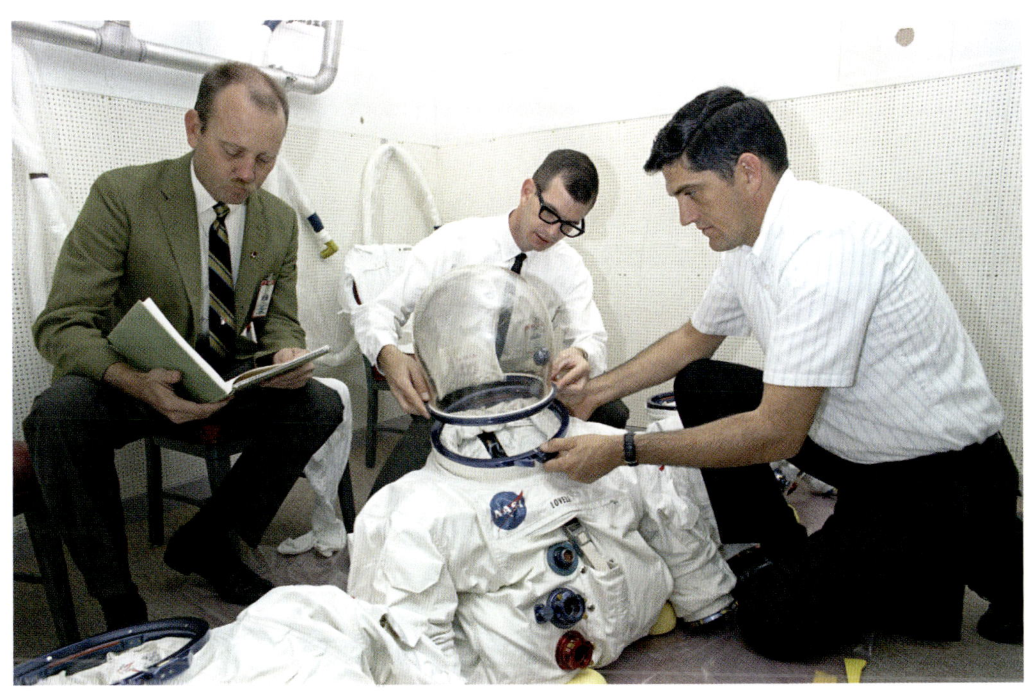

Sandridge and McDilda fit Lovell's helmet to its neck ring.

Left to right: NASA suit tech Joe Schmitt, McDilda, and Sandridge attach dual life vests to Borman, Lovell, and Anders.

McDilda and Sandridge assist Lovell and Anders (*hidden*).

Borman and Anders are suited as McDilda works with Lovell behind them, with Schmitt at right.

Borman, Lovell, and Anders

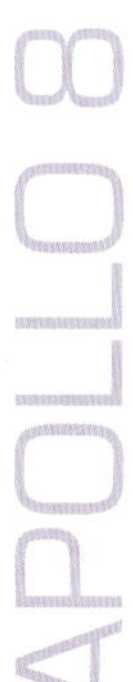

Borman, Lovell, and Anders in front of a CM mockup in the Flight Crew Training Building. The mockup was built and used for emergency egress training after the Apollo I fire almost two years earlier.

Borman

Lovell

The crewmen prepare for their crew portrait in front of the AMS.

The astronauts pose at the stairs to the AMS.

The crew during the photo shoot

Lovell

Anders

Borman

Alternate crew portrait

Anders enters the simulator first, with Borman and Lovell on the stairs behind him.

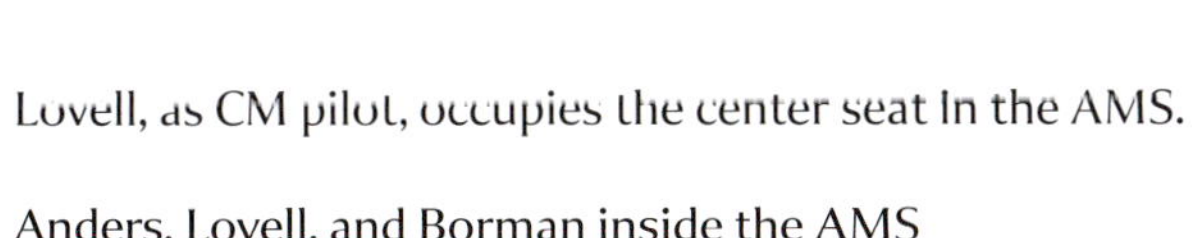

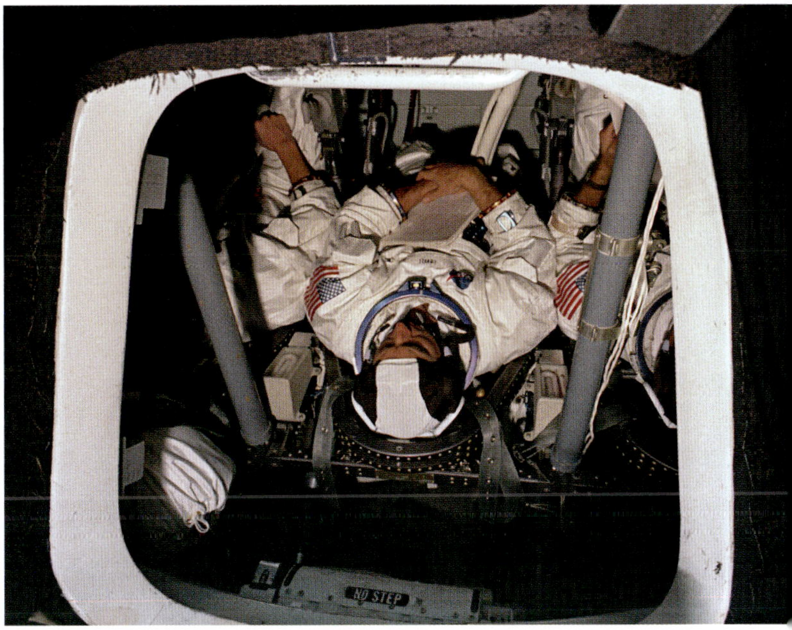

Lovell, as CM pilot, occupies the center seat in the AMS.

Anders, Lovell, and Borman inside the AMS

Anders, Lovell, and Borman hold a news conference at MSC's Building 2 auditorium on November 16, four days after their moon mission is announced. The astronauts say they're not upset about spending Christmas away from their families, with Borman adding that the whole program already keeps him away more than he'd like. MSC public affairs officer Paul Haney is almost out of frame (*at far right*).

"I don't think we're biting off more than we can chew," Borman says. "I'm not concerned at all."

Borman explains sun angles during lunar orbit insertion (LOI). "Frankly, I think I will have a very conservative approach," he says. "We won't burn the engine to go into lunar orbit unless we have a perfect spacecraft. If we have a problem . . . we'd be in trouble." One goal is to pass over planned lunar-landing sites during the same lighting conditions.

On November 18, personnel in Firing Room 1 in the LCC monitor Apollo 8 flight readiness test activities. *Left to right*: unidentified, test planning manager Bob Moser, unidentified, launch operations manager Paul Donnelly, lead test supervisor Gordon Turner, and chief test supervisor Bill Schick (*almost out of frame*).

Hans Gruene, chief of the Test Operations Office for Launch Operations, talks with Saturn program manager Lee James of MSFC.

Ike Rigell, KSC deputy engineering director for launch vehicle operations, is responsible for all Saturn V engineering personnel in the firing room during prelaunch preparations and countdowns.

Anders prepares to defend the mission by using lunar maps during a news briefing in the KSC Headquarters auditorium on the evening of November 20, after criticism the previous day from Sir Bernard Lovell, director of Britain's Jodrell Bank Observatory. The astronomer had called it "wasteful and silly" to risk astronauts when unmanned probes could provide the same data. Astronaut-geologist Jack Schmitt is seated in the background.

KSC public affairs chief Jack King (*right*) calls on a reporter. "We obviously don't agree with him," Anders says of the astronomer's comments. "We feel strongly that a manned platform in lunar orbit can add significant bits and pieces to scientific knowledge."

Cameramen for the three national TV networks film the briefing. *Left to right*: Wendell Hoffman (CBS), unidentified (NBC), and Chuck Pharris (ABC). CBS producer Jeff Gralnick sits to the left of Hoffman. ABC News science editor Jules Bergman is in the front row (*center*), with Schmitt now at right. Reporters seated on the far side include NBC's Jay Barbree and the AP's Howard Benedict.

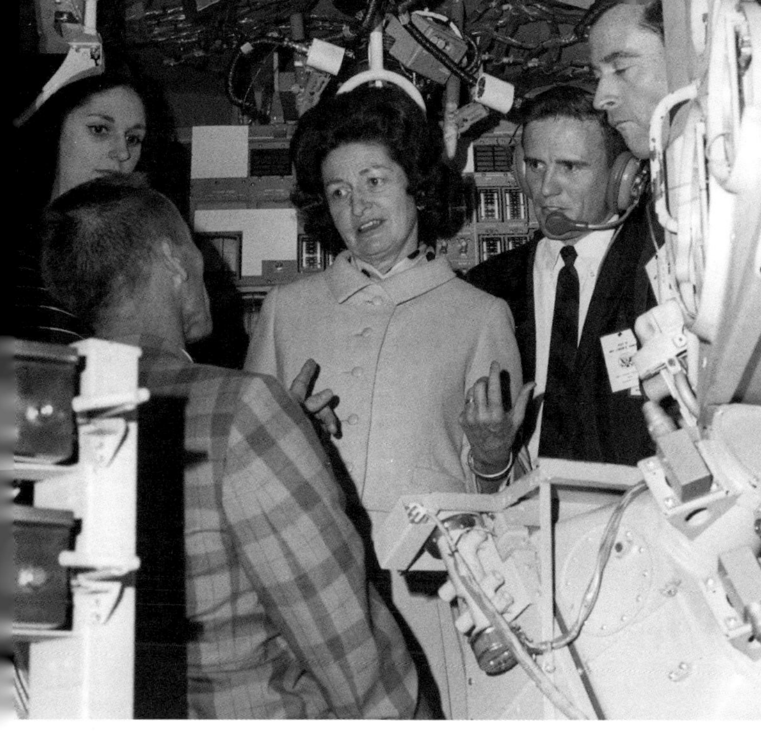

On November 23, astronaut Walt Cunningham (*left*) and John Young (*right*) brief First Lady Lady Bird Johnson on the LM simulator in the Flight Crew Training Building. Her daughter Lynda Robb listens behind Cunningham. The visitors make one successful "landing," but a second try is 200 feet off. "It was very exciting," Mrs. Johnson says.

Petrone briefs a group including Lady Bird and Lynda Robb near Pad A during their five-hour visit to KSC. Apollo 8 in the background elicits a "Wow!" from Mrs. Johnson. It's one stop for three buses taking guests and officials on the tour.

Below: NASA acting administrator Tom Paine presents Lady Bird with a quarter-scale LM model in the VAB for the LBJ Presidential Museum in Austin, Texas. "The size of everything here staggers the Imagination," she says. It was the second stop on her four-day, 6,000-mile Discover America tour to highlight advances made in space, education, health, urban renewal and conservation during her husband's five years in office, ending in two months.

CHAPTER 8

December 7–20, 1968

The astronauts pose with their CM-shaped mission emblem, sketched out by Lovell in the back seat of a T-38 jet over Colorado as he and Borman flew back to Houston from NAR in Downey on August 10. NASA artist William Bradley did the final art.

Lovell, Borman, and Anders hold their preflight news conference in MSC's Building 2 auditorium on December 7. Asked for an earthly comparison of the risk, Borman says, "It's something like a [fighter pilot] combat tour in Vietnam. The risks we take are acceptable ones. We have to accept if we believe it's worthwhile. If I ever feel it's not worthwhile, then I'll quit." A Walter Hyatt CSM model is at right.

Armored 35 mm motion picture cameras to film the launch are in place near the northwest corner of Pad A on December 8, as preparations are made for the MSS rollback before the fueled, or "wet," portion of the Countdown Demonstration Test (CDDT).

The crawler slowly backs the MSS away from the ML. The boost protective cover surrounds the CM. The white room is in place at the end of service arm no. 9 (*at left*), with service arm no. 8 below it.

Waves crash on the Atlantic beach in the background behind Apollo 8 in this view from the MSS. A yellow flame deflector is parked at right.

The crawler slowly backs the MSS down the ramp from the launch vehicle.

The lights come on at Pad A in the late afternoon.

Sunset pad view looking west, about 5:40 p.m. EST

Fifty new xenon searchlights illuminate Apollo 8 that night.

Seated, left to right: Apollo 7 astronauts Walt Cunningham, Donn Eisele, and Wally Schirra and Apollo 8 astronauts Anders, Lovell, and Borman upstairs in the Treaty Room before a December 9 White House dinner honoring recently retired NASA administrator Jim Webb and the Apollo astronauts. *Standing, left to right*: the first solo aviator to cross the Atlantic nonstop, Charles Lindbergh; Lady Bird Johnson; President Lyndon Johnson; Webb; and Vice President Hubert Humphrey. *Bill Taub archive*

Left to right: Johnson, Webb, and Humphrey watch as Lindbergh signs a piece of White House stationery for the first lady. The astronauts will sign it as well. Lovell and Anders are seated at left. *Bill Taub archive*

Later that evening, after dinner downstairs in the State Dining Room, Johnson presents Webb with the Presidential Medal of Freedom for his "farsighted and forceful leadership." Twenty-seven Apollo astronauts attend, including Dick Gordon (*at far left*). *Bill Taub archive*

Launch managers monitor a rerun of the final nine hours of the wet portion of the Apollo 8 CDDT in Firing Room 1 in the LCC before dawn on December 10. It was halted the previous afternoon because of a leaky helium valve for the S-IVB on the LUT. *Top row, left to right*: Jack King, KSC public affairs chief and launch commentator; John Williams, KSC spacecraft operations director; Andy Pickett, mechanical and propulsion director; Walt Kapryan deputy launch director; Rocco Petrone, launch director; Lee James of MSFC, Saturn program manager; and Ike Rigell, KSC deputy engineering director for launch vehicle operations. *Bill Taub archive*

Lt. Gen. Sam Phillips, Apollo Program director, oversees the test from the Operations Management Room.

The Saturn's second stage vents LOX vapor after Apollo 8 is fueled for the second time during the wet portion of the CDDT, which is completed that afternoon.

The prime crew poses with Apollo 8 that evening. The astronauts wear two-piece Beta cloth in-flight garments.

The backup crew (Armstrong, Aldrin, and Haise) pose as well. Armstrong and Aldrin would make the first lunar landing in six months; Haise would see the Moon from 158 miles away as Apollo 13 looped around it.

Both crews together

The next morning, December 11, Borman prepares in the suit room for the unfueled, or "dry," portion of the four-day CDDT.

ILC suit tech Ron Woods helps Lovell with a glove liner. The techs wear masks with the crew in medical quarantine, since the spreading the flu is one concern.

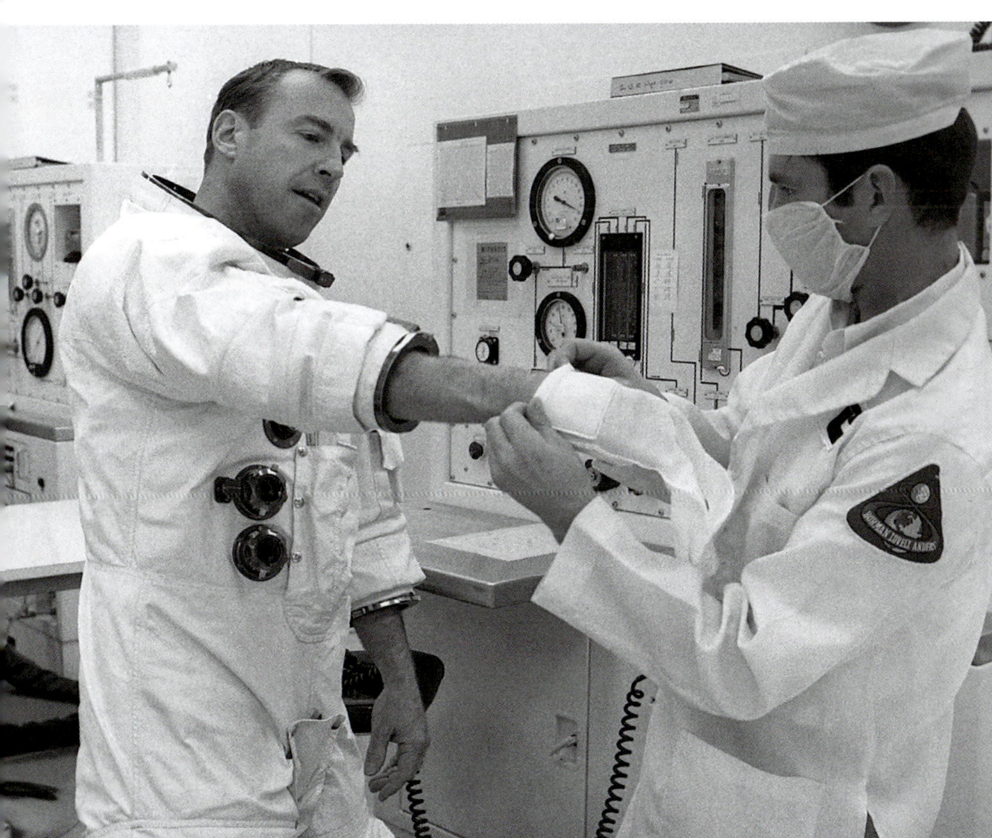

ILC tech Bob McDilda helps Anders don a glove. Chief astronaut Deke Slayton is at left.

Woods attaches Lovell's glove.

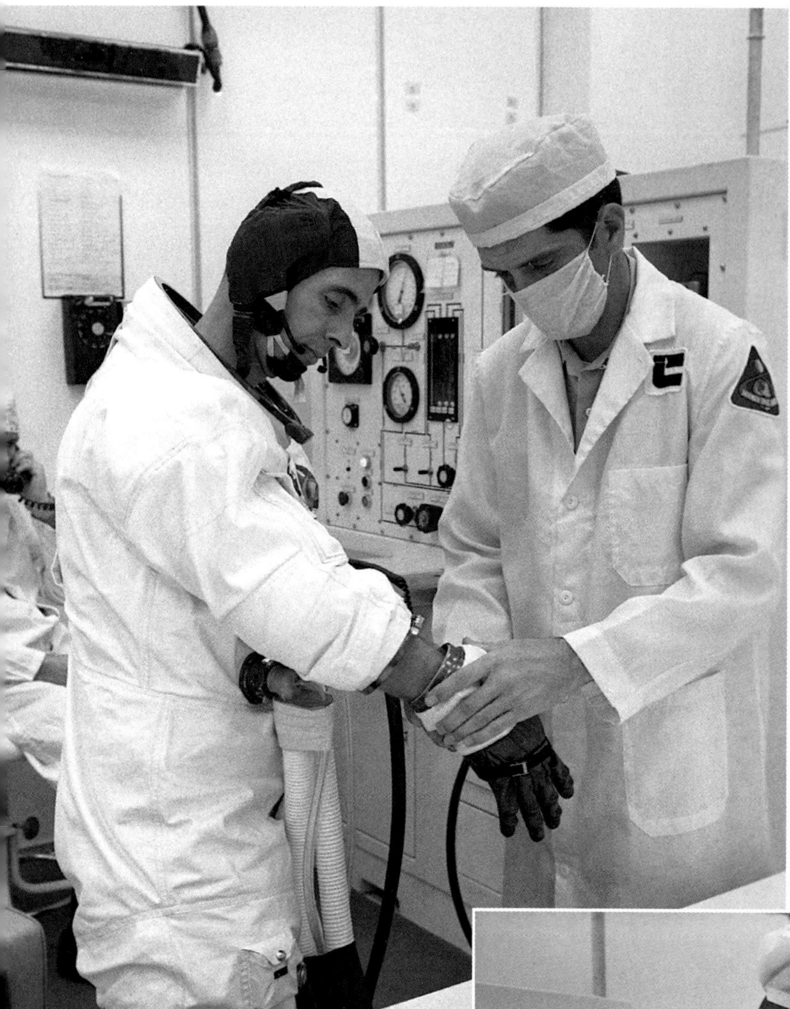

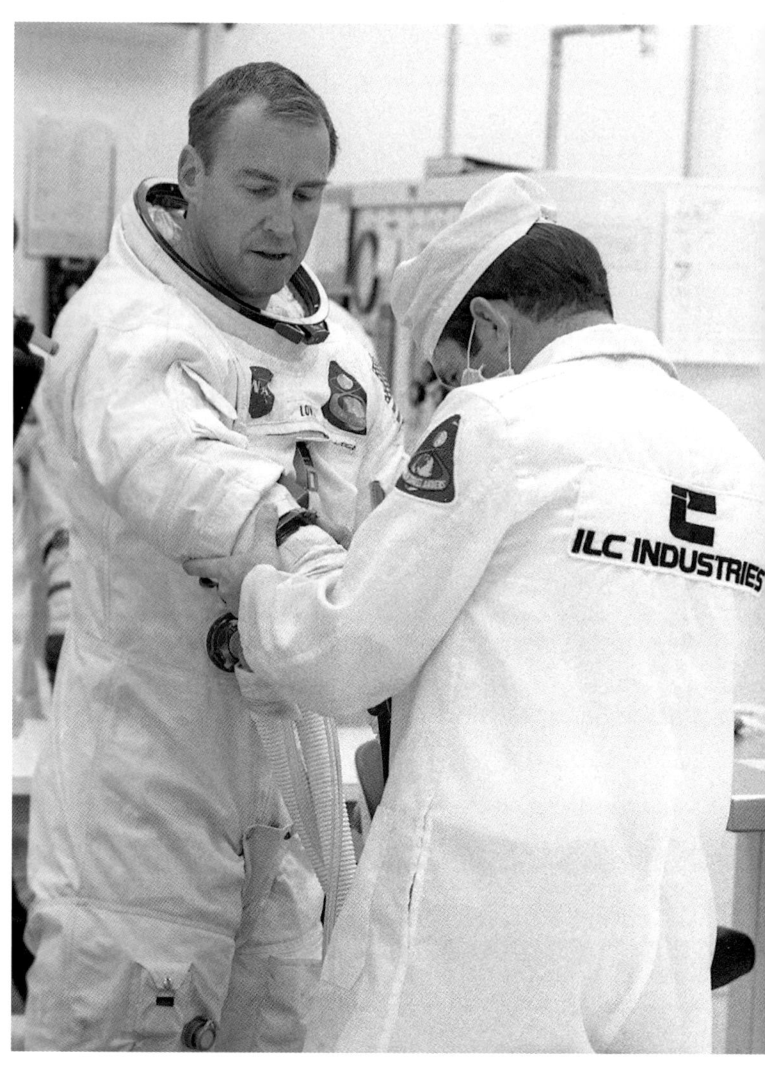

NASA suit tech Dick Sandridge attaches Borman's dual life vest.

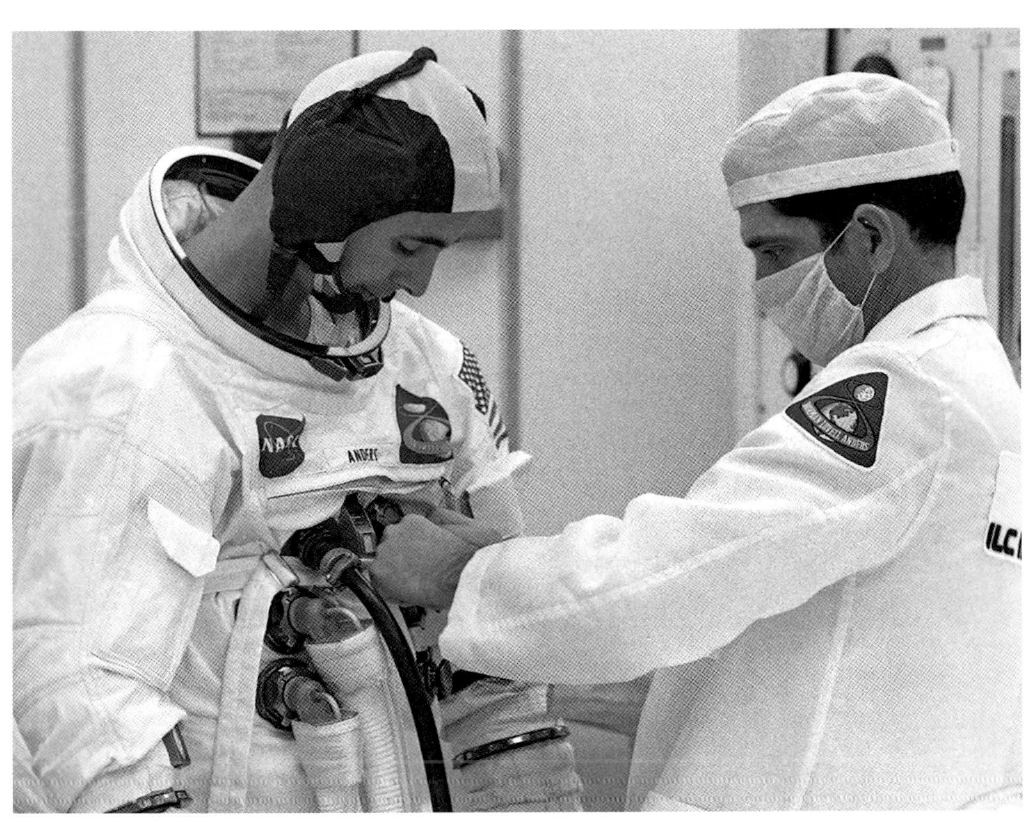

McDilda fastens Anders's life vest under his harness.

Sandridge (*out of frame at right*) fastens Borman's life vest under his harness.

Lovell and the other two astronauts begin to prebreathe pure oxygen to prevent the bends.

McDilda makes a final adjustment for Anders.

Anders is ready.

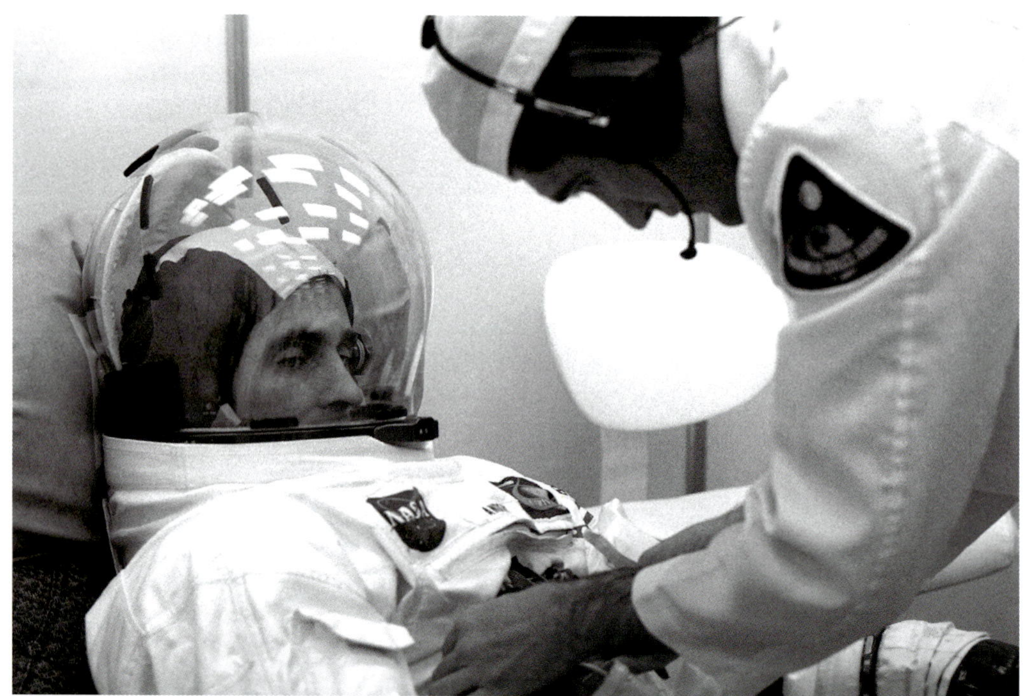

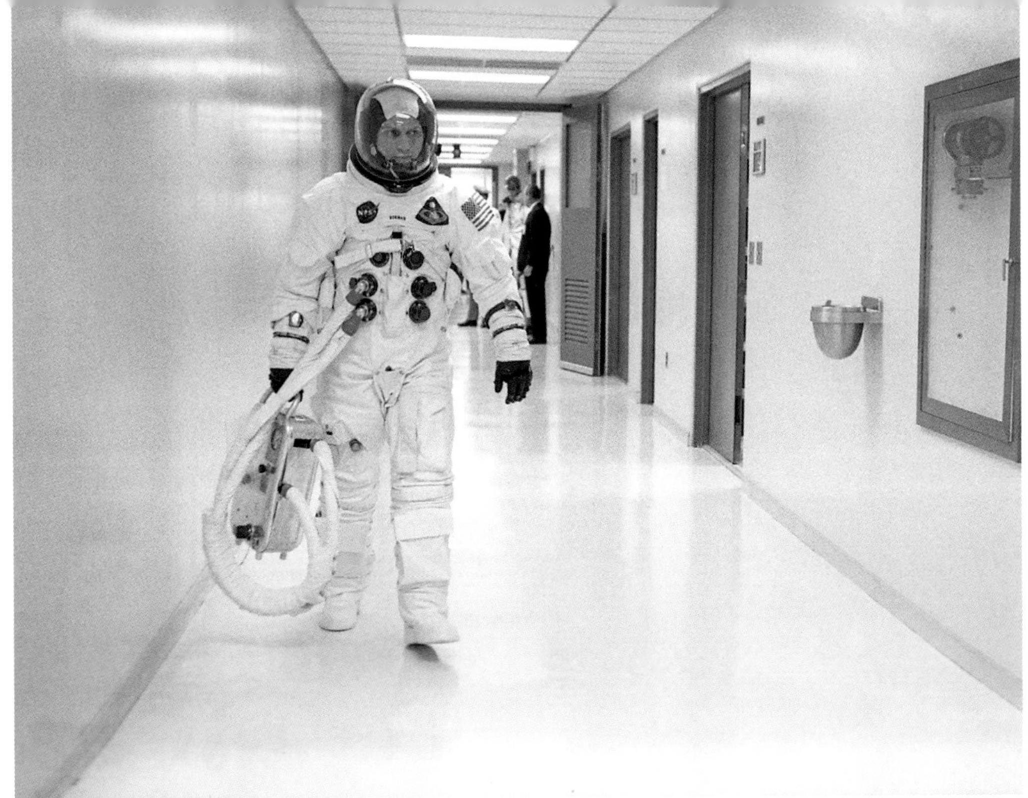

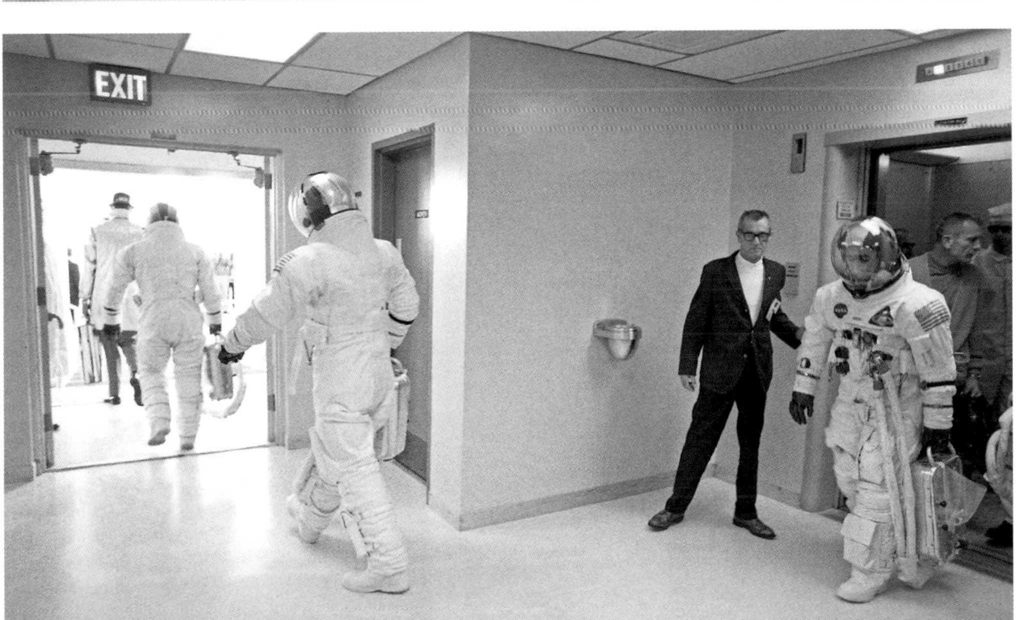

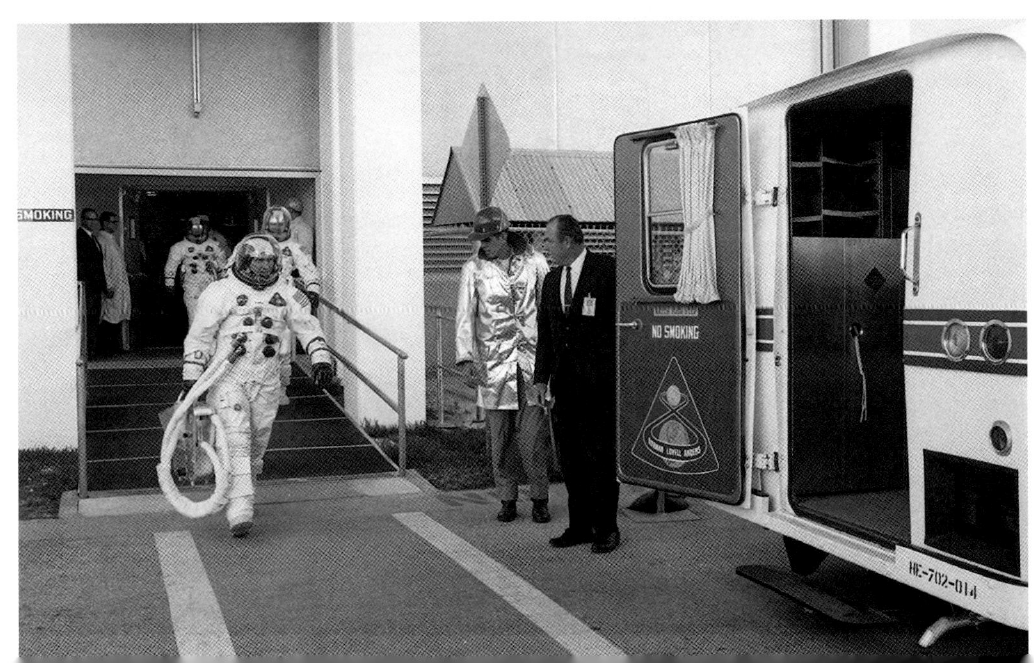

Borman heads for the third-floor MSOB elevator.

Borman (*left*) and Lovell head out the doors as Anders leaves the elevator. MSOB supervisor Tony Broadway is at center, with Slayton and suit techs following Anders.

Borman leads his crew to the Astronaut Transfer Van for the trip to Pad A. KSC security chief Charles Buckley holds the van door. The astronauts will spend more than five hours in the pressurized CM.

On December 17, Anders, Borman, and Lovell confer in the Flight Crew Training Building, with the AMS behind them. They train for six hours inside the simulator most weekdays, starting at 8:00 a.m.; the backup crew follows them in the afternoons.

Anders and Lovell share a laugh.

Anders on the stairs at the AMS

Lovell points to the Sea of Serenity (Apollo 15's eventual landing site) on a USAF moon map in the crew quarters conference room on December 19. *Left to right*: Dick Proffitt, spacecraft test conductor; Bill Schick, test supervisor; Paul Donnelly, launch operations manager; Lovell, Borman; Ray Roberts, launch vehicle test conductor; and Anders.

Lovell discusses a mosaic of Lunar Orbiter photos. *Left to right*: Borman, Roberts, Proffitt, Schick, Donnelly (*seated*), Lovell, and Anders. NASA photographer Bill Taub shoots in the background (*at left*).

Donnelly, Lovell, and Anders with the photo mosaic

A lone worker takes in the scene during a lighting test on December 19. The fifty new xenon searchlights, developed by the Duro-Test Corp. of North Bergen, New Jersey, duplicate the "color temperature" of sunlight for launch photography. The countdown had begun four nights earlier, with the clock starting at T-minus 103 hours.

Pad A, looking southwest during the lighting test

This view looks across the top of the flame trench, covered with FireX bricks. The engine-servicing platform is still in place on rails underneath the engine opening in the ML above. The crawler occupies this space when lifting or lowering the ML. Three hardened lights and a TV camera are on the trench edge.

Late winter afternoon views on December 20, looking west

View of Apollo 8 at about 5:45 p.m. Launch managers keep a close eye on the clouds for acceptable visibility during ascent the next morning.

Service arms no. 9 (with the white room) and no. 8 (SM) are in place as the MSS, with its clamshell enclosure for the CSM, begins to move away about 10:00 p.m.

The crawler slowly backs the MSS away.

View from the MSS as it moves away from the pad

Service lights on the MSS glow (*at left*), as the crawler moves it to its parking spot along the crawlerway.

Pad lights reveal fog that rolls in after sunset.

The V-shaped wind damper, attached to the LES, stabilizes the top of Apollo 8. Service arms nos. 9 (*hidden*), 8, and 7 are below it.

Fueling will begin before midnight as the countdown continues. The overnight low is 57 degrees F.

Lovell, Borman, and Anders in their crew quarters in the MSOB on December 20

The astronauts look over a photo of Hurricane Gladys in a spread on the Apollo 7 mission in the December 6 issue of *Life* magazine. They wear NASA-issued Omega Speedmaster watches.

The crewmen regard an aluminum Christmas tree in their quarters. They enjoy a steak dinner and go to bed early.

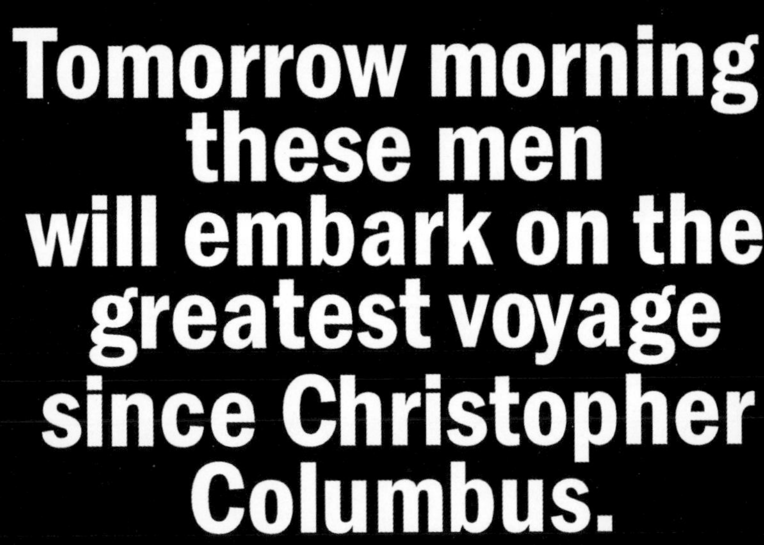

Quarter-page *New York Times* ad on December 20 for WCBS-TV

CHAPTER 9

December 21, 1968

The Saturn V's second and third stages vent LOX and liquid hydrogen in the early morning, with fueling underway. More than 783,000 gallons of the two cryogenics fill the booster's tanks, along with 203,400 gallons of RP-1 kerosene as first-stage fuel.

The launch team in Area C in the rear of Firing Room I in the LCC is busy just after 3:00 a.m. EST as the countdown continues. Consoles here monitor and record the status of various launch vehicle and pad systems. Four large Eidophor video projection screens are above them. The team encounters no significant problems during the count.

The crew has breakfast in the MSOB crew quarters conference room at 3:30 a.m. *At left*, Slayton and astronaut Jack Schmitt are almost out of frame. Borman, Lovell, and Anders sit across from them, joined by KSC security chief Charles Buckley and MSOB manager Hal Collins. The astronauts received a medical exam after waking up an hour earlier.

Anders finishes his orange juice, with Buckley next to him. The menu is filet mignon with eggs and toast, and coffee and juice. Armstrong also attends.

Borman talks with Apollo Spacecraft Program Office manager George Low, who had first proposed their lunar-orbit mission in August. Aldrin is next to Low.

Suit techs including NASA's Dick Sandridge (*right*) prepare the crew starting at 4:00 a.m. He stows a pen, mechanical pencil, and flashlight for Borman. Each crewman stands at an individual pressure garment assembly console.

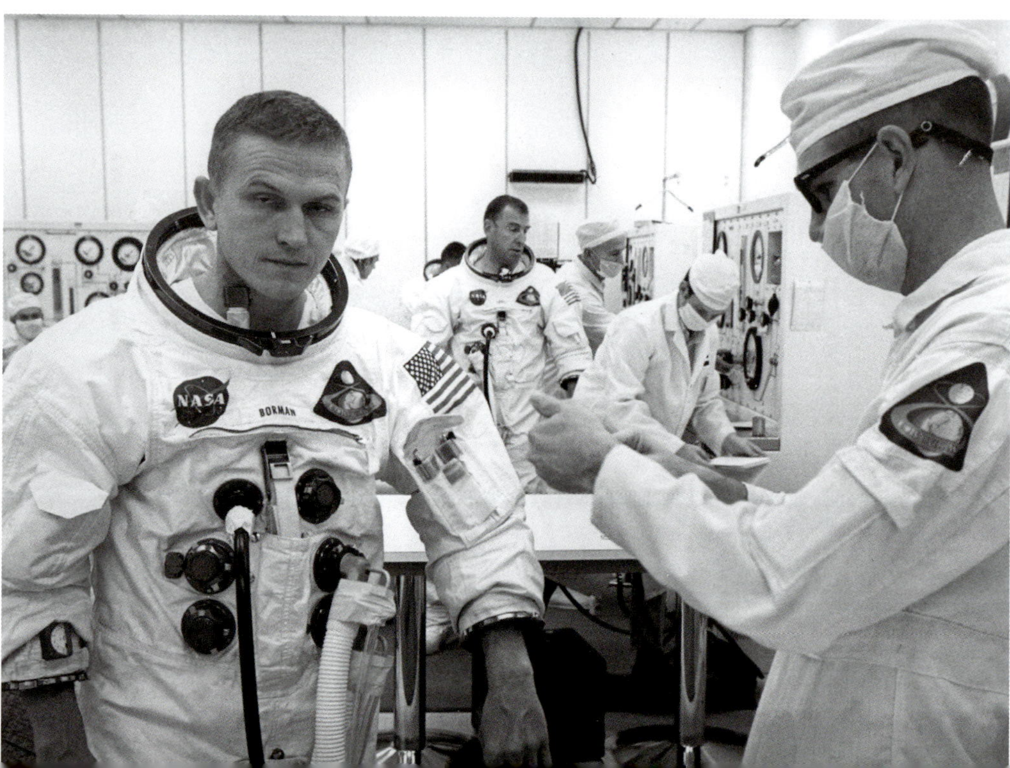

Lovell is assisted by ILC suit tech Ron Woods (*out of frame*).

ILC suit tech Bob McDilda works with Anders.

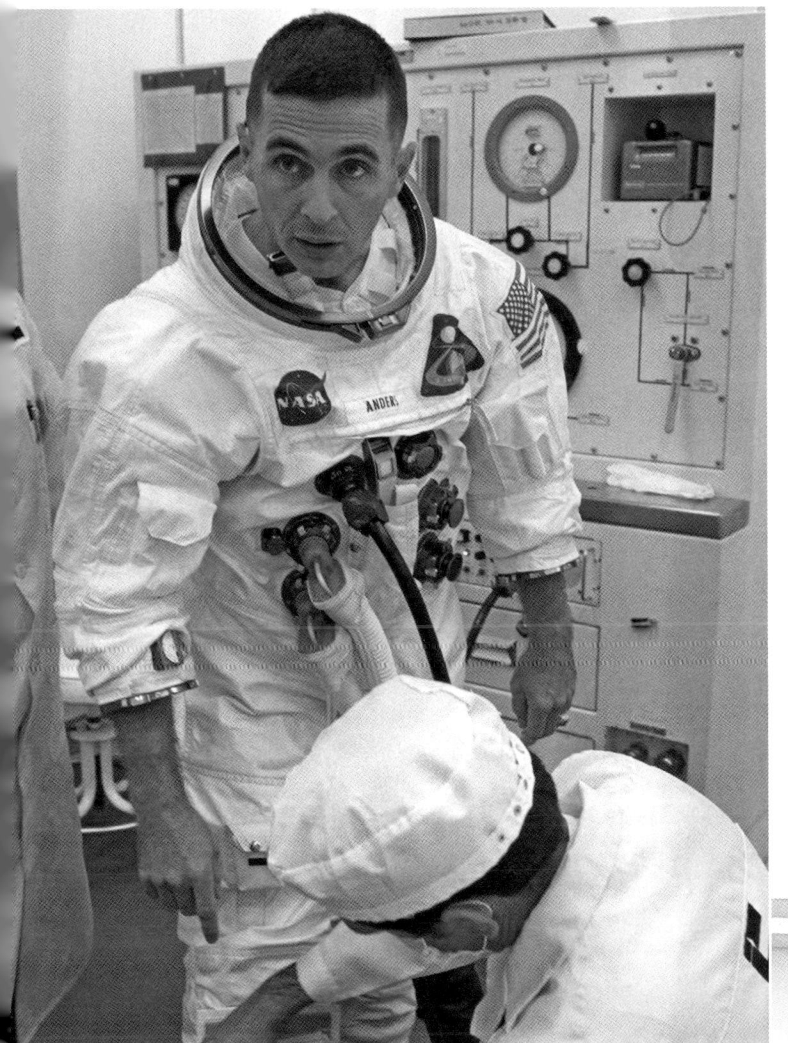

Borman is ready for Sandridge to help him don his communications carrier, gloves, and helmet. In the background, Woods closes Lovell's restraint and pressure-sealing zippers. Deke Slayton is at left.

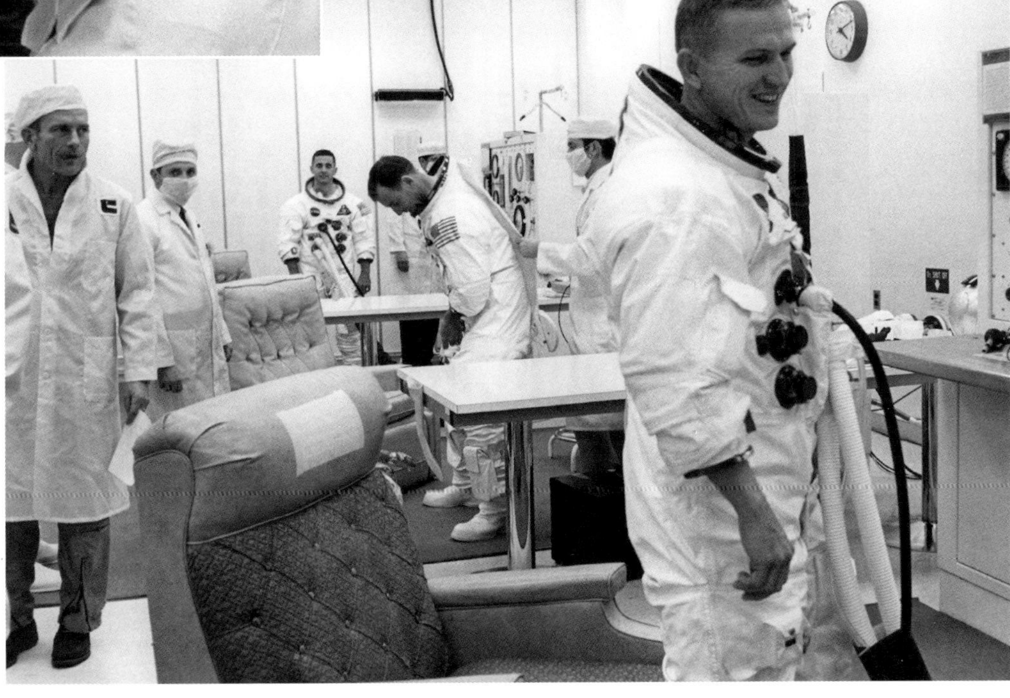

Borman reads a note about a gag Christmas stocking held by Sandridge. NASA suit tech Joe Schmitt is behind Borman.

Anders adjusts his communications carrier.

Schmitt and McDilda make adjustments to Anders's suit as he begins breathing pure oxygen.

The astronauts' suits are pressurized and checked for leaks. All three crewmen are ready to next transfer from the consoles' oxygen supply to portable ventilators.

Borman and Lovell lead a procession to the third-floor elevator.

Buckley holds the Astronaut Transfer Van door as Borman waves to a crowd of about five hundred news photographers and KSC workers. KSC firefighter Clay Crawford stands by.

Borman nears the van as Lovell waves, with Anders and Slayton behind them.

Borman steps in as Lovell and Anders follow.

Anders waves.

Photographers capture the van's departure at 4:32 a.m. for the twenty-minute trip to LC-39A.

Just before 5:00 a.m., a closeout crew member speaks to Anders in the white room as Borman (*left background*) prepares to enter the CM first; Anders will follow.

Lovell, who'd been waiting to enter the small white room, is the last to ingress, at 5:07 a.m. The CM hatch is closed at 5:34 a.m.

ABC News science editor Jules Bergman prepares for launch coverage in the ABC building at the LC-39 Press Site across the street from the VAB, south of the Saturn Causeway. A technician on a ladder adjusts a light (*at right*). The three TV networks will begin live coverage at 7:00 a.m. on this Saturday.

CBS Evening News anchor Walter Cronkite will serve as correspondent during launch coverage on the second floor of the CBS News building near ABC at the Press Site. He'll use the Earth and Moon models to demonstrate the distance between them. Producer Jeff Gralnick is at right.

Huntley-Brinkley Report coanchor David Brinkley will cover the launch for NBC News, sponsored by Gulf Oil Corp., in a trailer next door to CBS; the studio camera is from affiliate WFGA-TV in Jacksonville, Florida.

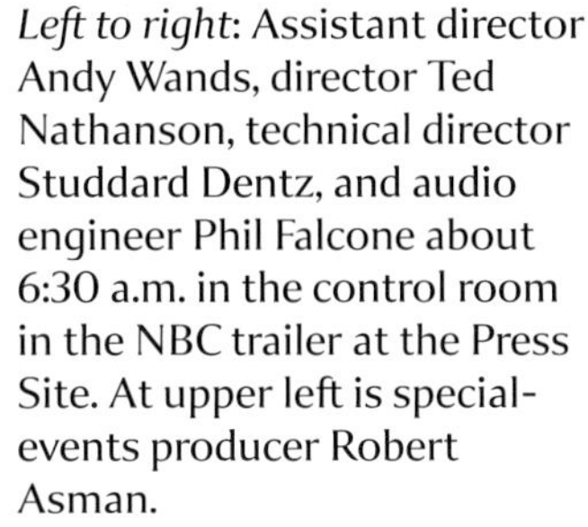

Left to right: Assistant director Andy Wands, director Ted Nathanson, technical director Studdard Dentz, and audio engineer Phil Falcone about 6:30 a.m. in the control room in the NBC trailer at the Press Site. At upper left is special-events producer Robert Asman.

Background to foreground: Falcone, Dentz, Nathanson, and Wands

A broadcaster waits in the WKKO-AM mobile unit from Cocoa Beach, Florida.

An NBC cameraman at the Press Site trains his GE PE-350 camera at Pad A on the cool morning.

A military Sea Stallion helicopter patrols in the vicinity. "We really have a beautiful morning for the flight to the Moon," says launch commentator Jack King in the firing room. "The weather conditions are very satisfactory for a launch attempt." *Photo by Jacques Tiziou*

Motion picture and still photographers at the Press Site prepare for launch, among the 1,356 accredited US and international news media from twenty-two countries, but none from the Soviet Union.

Directly behind the photographers in the previous photo are print and radio reporters in the grandstand.

Grandstand seats for journalists can be reserved in advance, with phone or broadcast lines installed as needed.

A photographer checks on one of a pair of Nikon F cameras fitted with motor drives and an 800 mm telephoto lens, and a 500 mm mirror lens (*top*).

Former NASA administrator Jim Webb talks with Joan Kennedy and her husband, Sen. Ted Kennedy (D-Mass.), in the VIP viewing stands near the VAB. It is Kennedy's first launch and Webb's first Saturn V launch.

Buzz Aldrin chats with a guest in the viewing stands at the VIP stands. President Johnson and Vice President Humphrey, both sick with the flu, are unable to attend, but two Supreme Court justices and members of the House and Senate are there. Also present are aviation pioneers Charles Lindbergh, Chuck Yeager, and Jackie Cochran.

Florida governor Claude Kirk (R) (*center*) at the VIP site claims he attends all manned launches "because they're all Republican shots."

Spectators wait at Jetty Park at Port Canaveral, 13 miles south of LC-39. The day's launch window extends from 7:51 a.m. to 12:31 p.m.

Eastbound morning traffic flows through a toll plaza on the Bennett Causeway on Florida State Road 528, as other cars park on the north side, facing the launch. The Canaveral Barge Canal, used to bring Saturn stages from Port Canaveral, is at top left.

Apollo 8's first stage vents LOX just before ignition. The spacecraft and launch vehicle are now on internal power, and the final three minutes of the countdown are under the control of an automatic checkout sequencer.

A TV newsman (*center*) records his report with the launch seconds away. The morning has warmed up to 60 degrees F, still the lowest temperature for any Saturn launch.

The five F-1 engines of the first stage ignite sequentially in less than two seconds at nine seconds to launch. "We have ignition sequence start; the engines are on. Four . . . three . . . ," says King. Water pours into the flame trench.

More than 1,200 pounds of frost fall from the Saturn V's skin as the first stage quickly builds up 7.5 million pounds of thrust before the 6.2-million-pound booster can be released. "Two . . . one . . . zero . . . , we have commit!" King announces. The remaining service arms and three tail service masts retract, and the four hold-down arms release the behemoth.

DECEMBER 21, 1963

Apollo 8 is illuminated as flames and smoke billow from beneath the ML. This view from the VAB roof looks across East Creek and Broadaxe Creek, with the Atlantic Ocean in the background.

Left to right: Webb, the Kennedys, and the first NASA administrator, T. Keith Glennan, watch the liftoff. "I'm very hopeful it will be the success it's started out to be," Kennedy says later. "It all happened so quickly," Joan adds. "I'd like to see it again in slow motion. I'm going to Palm Beach and tell President Kennedy's parents all about it."

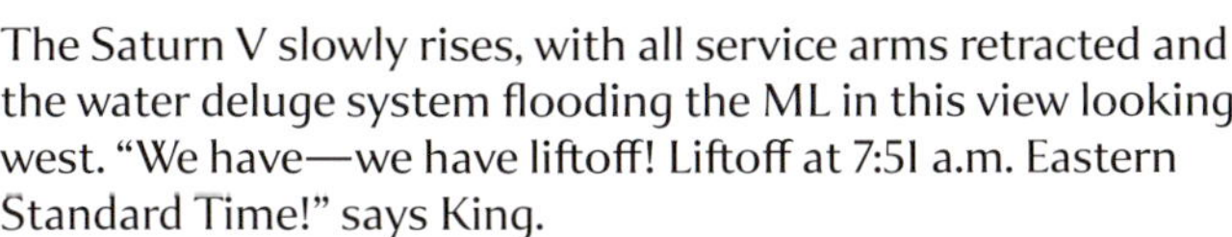

The Saturn V slowly rises, with all service arms retracted and the water deluge system flooding the ML in this view looking west. “We have—we have liftoff! Liftoff at 7:51 a.m. Eastern Standard Time!” says King.

“Liftoff . . . the clock is running,” Borman radios simultaneously.

Apollo 8 five seconds after launch. The Saturn executes a slight yaw to avoid any LUT impact.

A simultaneous view, looking southwest

Apollo 8 clears the LUT after nine seconds, and Mission Operations Control Room (MOCR) 2 in Houston assumes flight control. "Roll-and-pitch program," Borman confirms, as the Saturn's guidance system rolls the booster to the correct flight azimuth over the Atlantic.

Apollo 8 climbs on a pillar of flame and smoke. "How do you hear, Houston?" Borman soon asks. "Loud and clear," responds capcom Mike Collins in the MOCR.

Photographers at the Press Site follow the ascent with a few cirrus clouds and about 10 miles of visibility. *Photo by Jacques Tiziou*

Apollo 8 leaves a trail of fire against the morning Florida sky.

Director of flight operations Chris Kraft (*left*), MSC director Robert Gilruth, and deputy director George Trimble monitor the launch from their consoles in MOCR 2 on the third floor of the MCC. Mission commentator Paul Haney (*headset*) is at the next console.

Flight director Cliff Charlesworth follows ascent at his console in the MOCR during his Green team's eight-hour shift. Milton Windler will lead the Maroon team, followed by Glynn Lunney's Black team of controllers.

"Apollo 8, you're looking good," radios Collins. *AP photo*

A vapor cone forms around the second stage at one minute after launch as the Saturn approaches Mach 1 at more than 4 miles downrange. Maximum dynamic pressure on the vehicle occurs nineteen seconds later.

Apollo 8 is obscured by exhaust as it pitches over and heads southeast toward the northwestern coast of Africa on its way to Earth orbit.

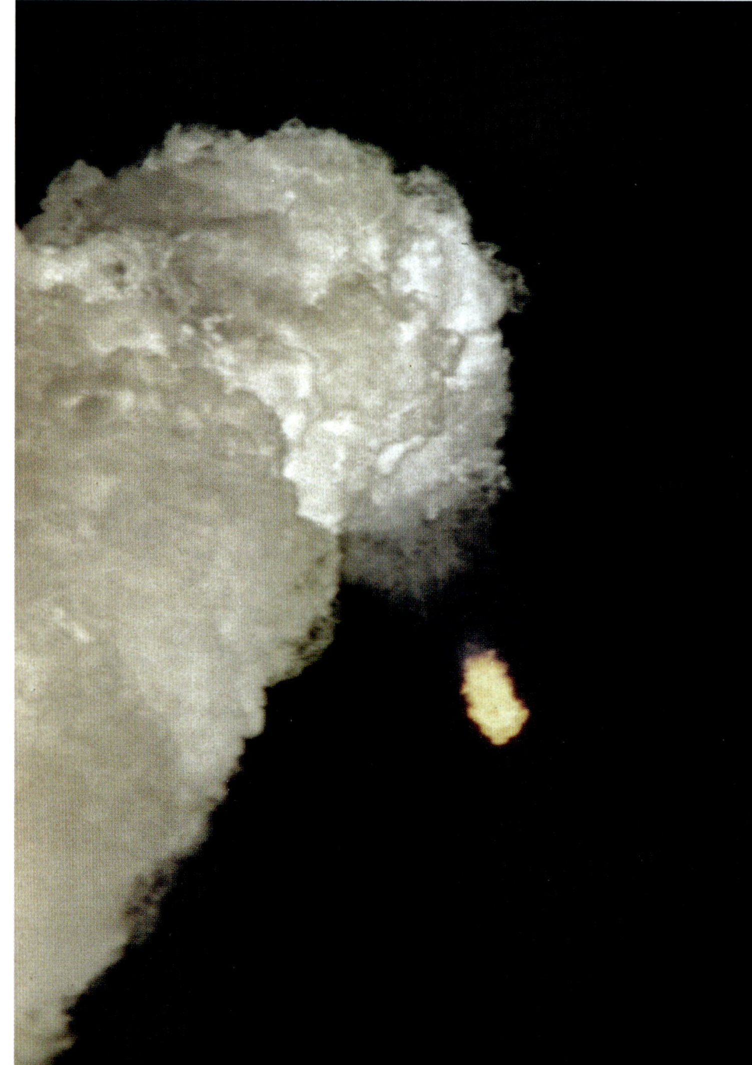

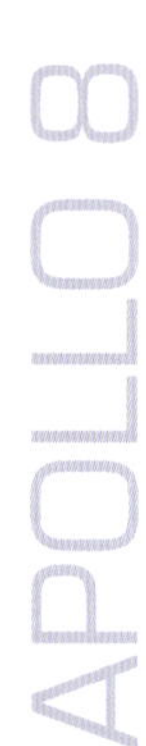

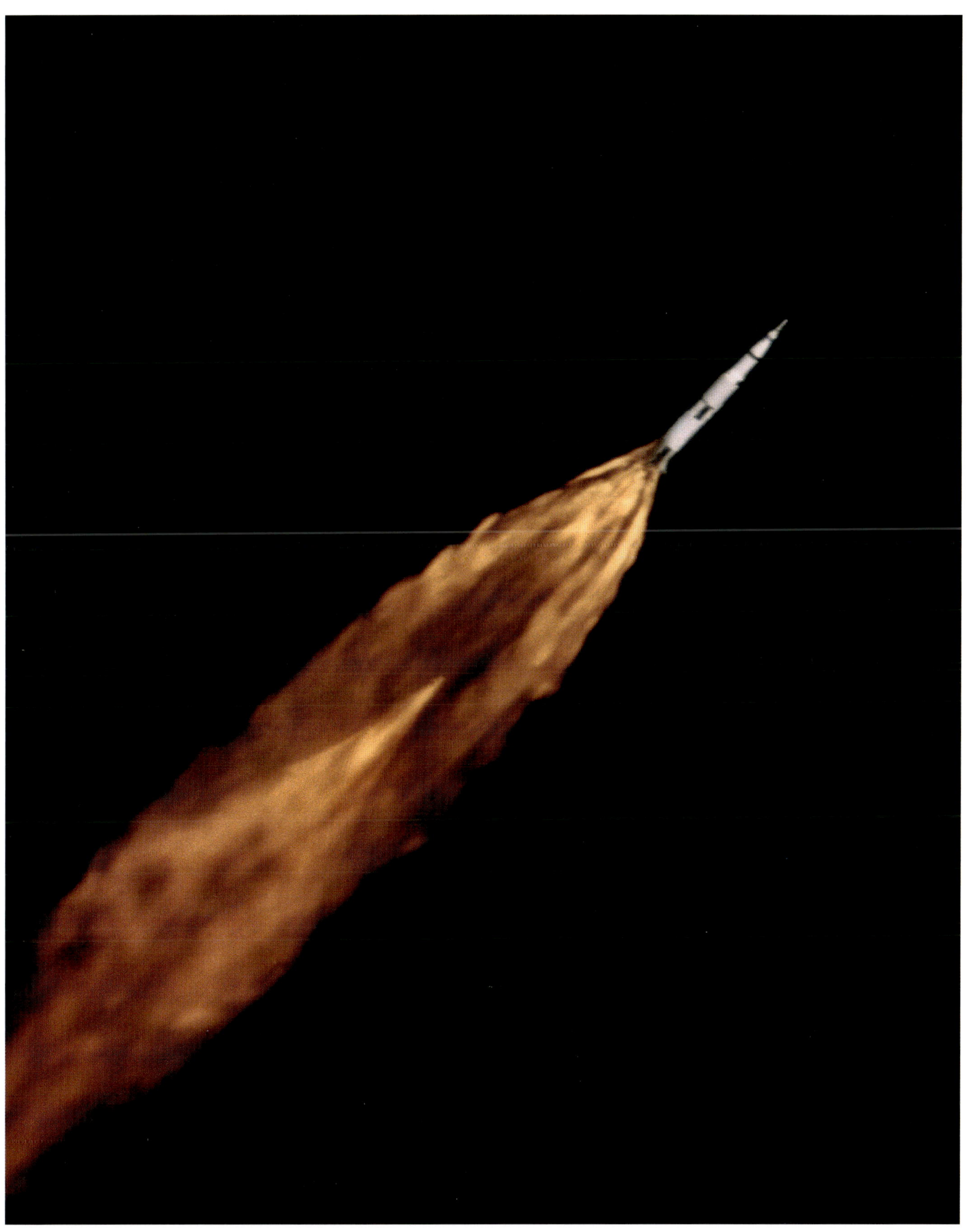

Apollo 8 streaks through the morning sky. The Saturn V's first-stage engines burn for two minutes and thirty-three seconds.

Engineers in the separate Recovery Room near the MOCR monitor the ascent and the status of Atlantic ships.

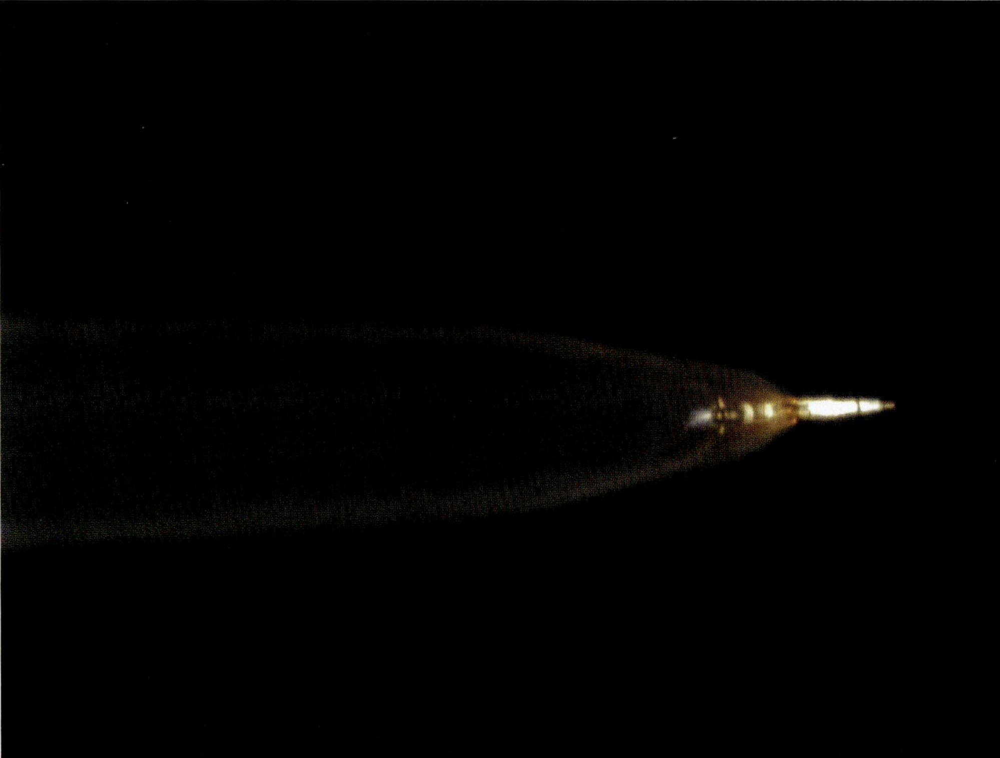

The first-stage engines still glow at two minutes and thirty-five seconds into the flight in this tracking camera photo as the five engines in the second stage ignite. "Staging!" Borman radios. Apollo 8 is now 40 miles in altitude and 55 miles downrange.

Spectators in Jetty Park watch until the booster is out of sight. Area crowds are estimated at 250,000, the largest for a launch, and postlaunch traffic is a problem until late afternoon.

Left to right: Lovell's children Jay (age thirteen), Jeffrey (three), and Susan (ten); his wife, Marilyn; and daughter Barbara (fifteen) watch from near the VAB. Wives Susan Borman and Valerie Anders watch on TV in their homes in Seabrook, Texas.

Journalists at the grandstand watch the launch in the winter morning sunlight.

Motion picture cameras on motorized tracking pedestals with telephoto lenses follow the ascent.

Oswaldo López Arellano, president of Honduras (*center*), watches the launch at the VIP site. He had come to Miami for his annual physical and a family vacation. Behind him are Mrs. Tom Paine (*left*) and Mrs. Kurt Debus. At left is Daniel Callahan, KSC deputy director of administration.

Photographers break down their equipment at the Press Site. *Photo by Jacques Tiziou*

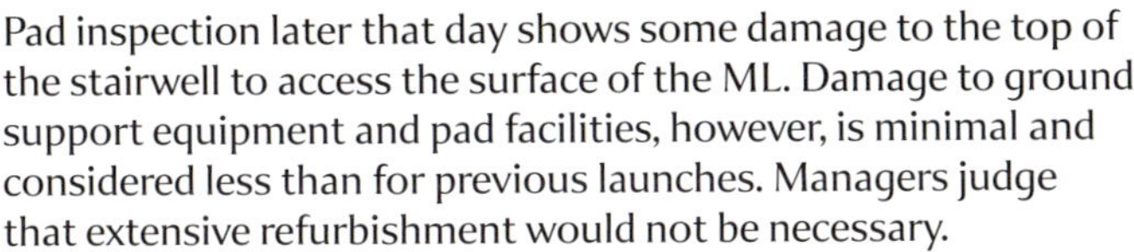

Pad inspection later that day shows some damage to the top of the stairwell to access the surface of the ML. Damage to ground support equipment and pad facilities, however, is minimal and considered less than for previous launches. Managers judge that extensive refurbishment would not be necessary.

A video camera dangles from its cable on the side of the white room.

Launch forces cracked the lens of this video camera.

CHAPTER 10

December 21–23, 1968

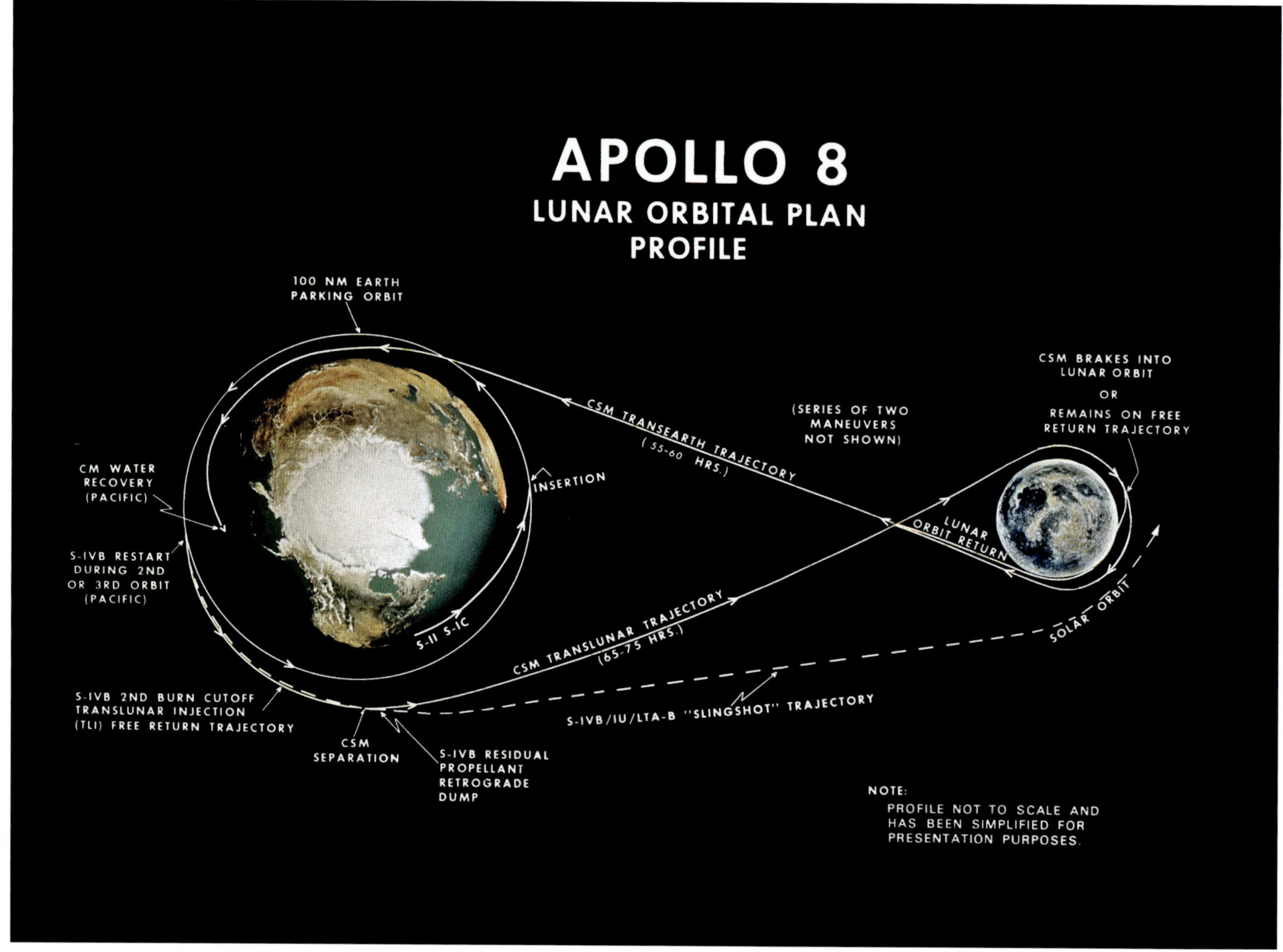

A NASA illustration shows the flight's "figure eight" path, which is the most time- and fuel-efficient because of orbital mechanics involving the Moon's gravity.

The S-IVB is in Earth orbit in art by McDonnell Douglas Astronautics Co., prime contractor for the third stage. Its J-2 engine has fired for just more than two minutes to put Apollo 8 into a parking orbit about 115 miles in altitude at 8:02 a.m. EST. The spacecraft will make one and a half orbits before the S-IVB ignites again over Hawaii.

Green team flight controllers in MOCR 2's first row, known as "the trench," prepare for the S-IVB's engine to reignite for translunar injection (TLI). The consoles include booster systems, flight dynamics, and guidance. At two hours and twenty-seven minutes into the flight, Capcom Mike Collins tells the crew, "Apollo 8, you are go for TLI."

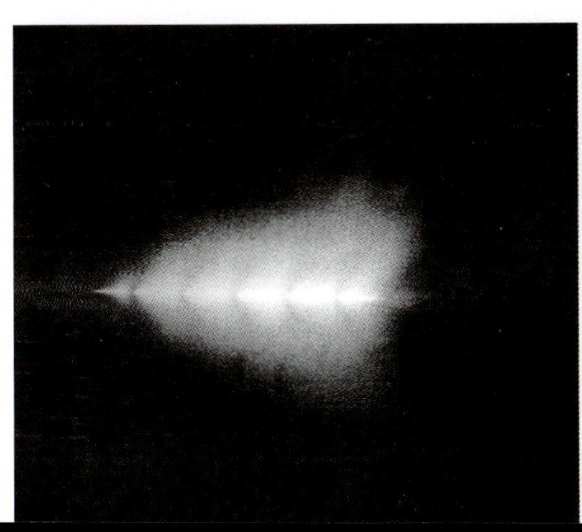

A satellite tracking camera at the Smithsonian Astrophysical Observing Station on Maui, Hawaii, captures two pictures of the burn starting at 4:41 a.m. Hawaii Standard Time. The firing is also visible from the ground. *Photos by David Le Conte, Joe Coldwell, and Bill Perry / Smithsonian Astrophysical Laboratory*

Left to right: Kraft, Gilruth, and Trimble watch the five-minute SPS burn in the top (senior management) row of the MOCR. "Your cutoff looked very good down here," Collins tells the crew. "We have a whole room full of people that say you look good." Kraft says out loud, "You're really on your way now."

NAR art illustrates the CSM and third stage leaving Earth orbit, reaching about 24,200 miles per hour. The trajectory, computed from the Saturn's instrument unit, provides a "free return" to Earth around the Moon as an option, and Apollo 8's speed will gradually slow as it coasts.

This first photograph from the spacecraft, taken with a 70 mm Hasselblad camera, shows the US Atlantic coast (*at left*), from the Chesapeake Bay to Florida. "Tell Conrad he lost his record," Borman says, a reference to Pete Conrad and Dick Gordon's 1966 Gemini XI flight, which reached 850 miles in altitude.

The SLA's four 29-foot panels, which enclosed LM Test Article (LTA)-B during launch, are jettisoned using springs in NAR art. They had remained attached during Apollo 7, but after one of its panels initially failed to fully open, the new procedure was implemented.

NAR art shows transposition at three hours and twenty-one minutes into the flight so the crew can photograph the stage. It also serves as practice for subsequent missions that will need to dock with an LM. The high-gain antenna has been deployed at the aft end of the SM.

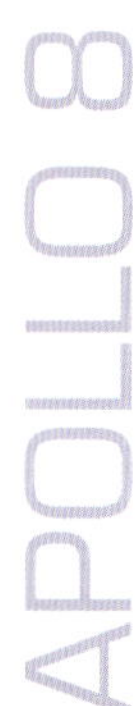

"What a view!" says Borman. LTA-B served as ballast to simulate the mass of an LM; the SLA panels have drifted away.

The stage is less than 1,000 feet away as it vents remaining fuel, which makes Borman uncomfortable.

The S-IVB appears to be closing in on the CSM, and flight controllers advise the crew to maneuver "up" to get away from the stage.

The CSM uses its small reaction control system (RCS) thrusters to leave the stage behind and depart for the Moon at four hours and forty-five minutes into the flight. The S-IVB is sent on a trajectory to fly past the Moon in three days and enter solar orbit. *Illustration by Gary Meyer*

Marilyn Lovell holds son Jeffrey as she talks with reporters in Cocoa Beach that afternoon. "I think it's a real treat to be able to watch from the Cape," she says. "My feelings at liftoff were rather incredible. You can't describe how it feels to see your husband sitting atop such a big rocket. I wouldn't trade it for anything, and I hope Jeffrey won't ever forget it." *UPI photo*

Valerie Anders (*left*) and Susan Borman are pleased after watching the launch and first hours of the mission on TV at their homes in Seabrook, near MSC. *AP photo*

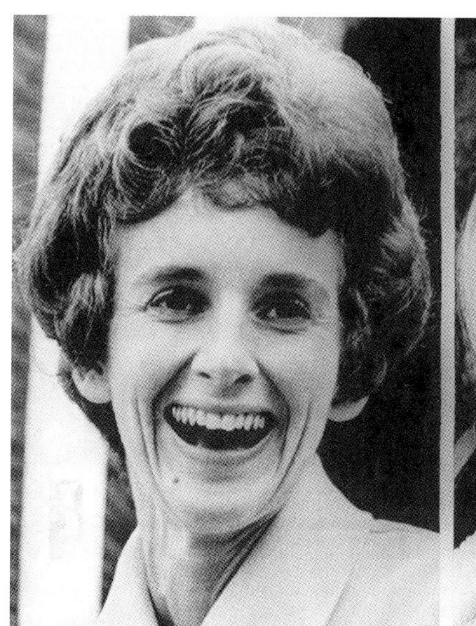

The third DSN antenna (HSK) is southwest of Canberra, Australia. Each site has multiple dishes, and the complete network also provides radar and radio astronomy observations.

Following TLI, Apollo 8 begins using the DSN. The crewmen had conducted voice checks through the Goldstone station (GDS), northeast of Barstow, California, during their first orbit.

The DSN antenna west of Madrid, Spain (MAD). Each site includes a 230-foot-diameter steerable antenna located in semimountainous, bowl-shaped terrain to help shield against radio frequency interference.

During the Mercury and Gemini programs, NASA communicated with orbiting astronauts by using a worldwide network of ground tracking stations, supplemented with ships and aircraft, known as the Manned Space Flight Network.

With Apollo traveling to the Moon, however, Mission Control needed a new technique both to track the spacecraft at such distances and to allow continuous two-way communication. The result was the Unified S-Band System, which combined tracking, voice, data, telemetry, and television into a single microwave carrier. It used three huge antennas spaced about 120 degrees apart around Earth, known as the Deep Space Network (DSN), operated by NASA's Jet Propulsion Laboratory in Pasadena, California. The array also communicates with interplanetary and interstellar probes.

The first photo of the entire Earth taken by a human, likely Anders, almost five hours into the flight at a distance of almost 20,000 miles. South America is at center, with the west coast of Africa at top left.

The next day, December 22, the Black team's flight surgeon, Dr. John Zieglschmid (*bow tie*), and chief flight surgeon, Dr. Charles Berry, and incoming Green team's flight surgeon, Dr. Willard Hawkins (*right*), discuss the crew's medical issues. Borman reported flu-like symptoms; all three had suffered brief bouts of nausea but now say they're feeling much better. Chief astronaut Alan Shepard is farther down, with incoming Maroon team capcom Ken Mattingly behind him. Capcom Collins is next to Shepard.

NASA first experimented with television from orbit during Mercury astronaut Gordon Cooper's 1963 flight. He briefly transmitted a few slow-scan black-and-white TV images at a rate of one frame every two seconds, with a resolution of 320 lines.

For Apollo, RCA built six 4.5-pound vidicon TV cameras that produced ten frames per second at 320 lines. RCA also provided ground equipment to convert the transmissions for domestic broadcast. They used a magnetic disc recorder similar to those used for sports instant replay to convert the slow-scan signal to the US broadcast standard of 525 lines at thirty frames per second. That was sent to MSC on landlines, which provided the transmissions to the three national American networks. Europe and East Asia received the video relayed by NASA satellites over the Atlantic and Pacific.

During Apollo 7, only two tracking stations had the converters, but all three DSN stations had them for Apollo 8 (the TV signal was multiplexed into the CSM's new Unified S-Band System). Goldstone sent the video to MSC on landlines; the only Madrid TV reception was relayed to MSC by satellite.

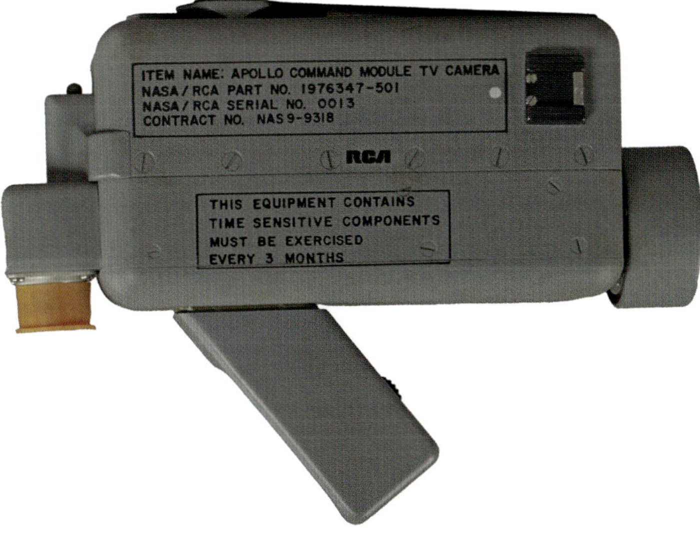

An RCA black-and-white vidicon TV camera identical to Apollo 7's is carried on Apollo 8. Two lenses are also provided: a 160-degree wide-angle lens and a 9-degree 100 mm telephoto lens. *Smithsonian National Air and Space Museum*

Sitting at the EECOM console, right to left: Maroon team flight controllers Joe DeAtkine, Clint Burton, Charles Dumis, and Buck Willoughby. *Standing, right to left*: Tom Hanchett, Sy Liebergot, backup LM pilot Fred Haise, Jack Hammond, Merlin Merritt, Dick Gover, and Rod Loe.

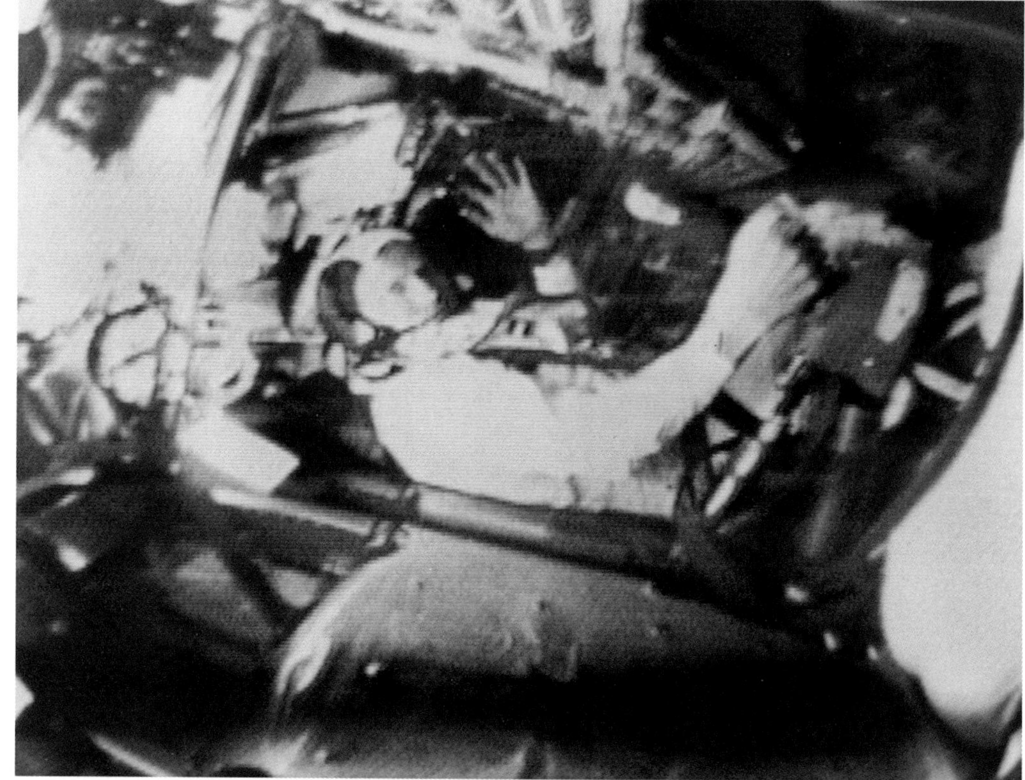

Borman waves at the beginning of the crew's first TV transmission at thirty-one hours into the mission (about 3:00 p.m. CST) as Anders holds the camera. The downlink is received through Goldstone.

Anders shows Earth, more than 120,000 miles away, but the contrast is too bright for the camera, and the window is hazy. Ice from vented water causes speckling. "It's a beautiful, beautiful view with a predominantly blue background and just huge covers of white clouds," says Lovell.

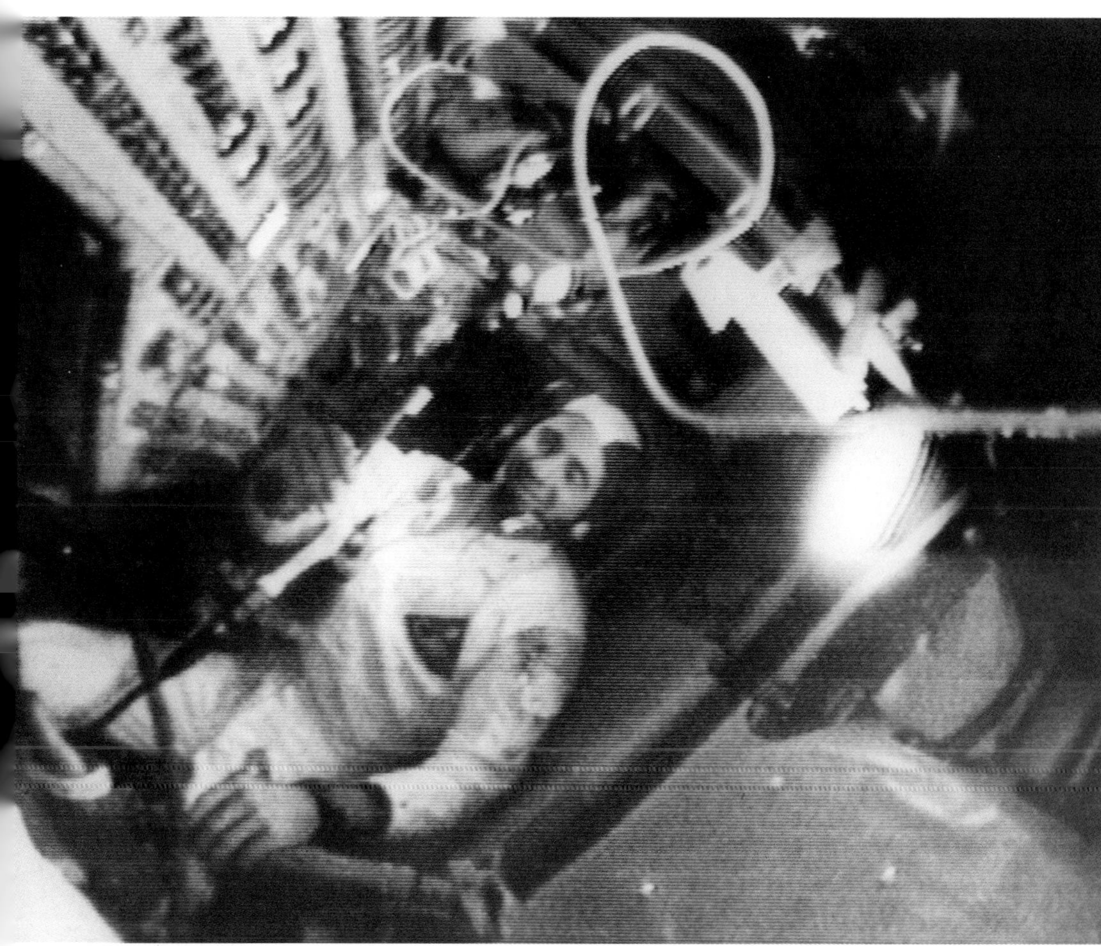

Anders holds his toothbrush. “This transmission is coming to you approximately halfway between the Moon and Earth,” notes Borman.

Left to right: Phillips, Low, Gilruth, and Trimble follow the flight on December 22.

Anders holds the TV camera's telephoto lens, which does not seem to work.

Lovell wishes his mother, Mrs. Blanche Lovell, a happy birthday. The spacecraft is approximately 120,650 miles from Earth and moving at 3,207 mph.

Borman waves to close out the first TV transmission after just more than ten minutes. "Goodbye from Apollo 8," he says.

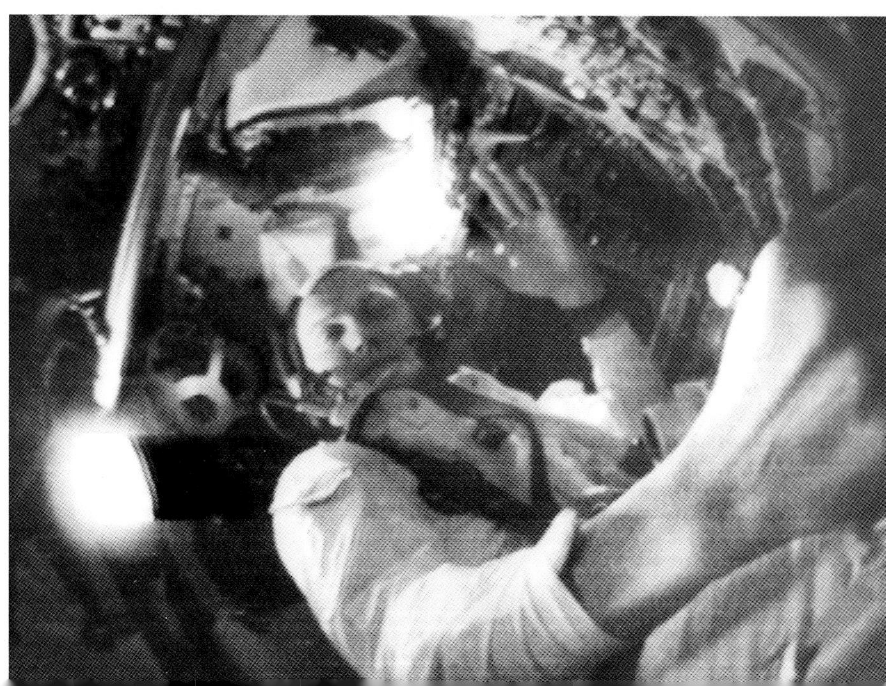

Greg Anders, six, looks at a photo of his dad with Valerie Anders from the second telecast on December 22 at their home in Seabrook. *AP photo*

Borman is in his couch, with Lovell in the lower equipment bay, as they coast toward the Moon (16 mm still frame).

Lovell, serving as navigator, sights through the telescope at the Guidance, Navigation, and Control Station to help align the CM's inertial measurement unit. The flight makes the first significant use of the Apollo guidance system (16 mm still frame).

Lovell holds a flashlight. "He has really outdistanced us all in the beard race," says Borman. "Jim has quite a beard going already." The eyepieces for the sextant and telescope are at right (16 mm still frame).

On December 23, flight controllers watch the second TV transmission, starting at fifty-five hours into the mission (about 3:00 p.m. CST). Anders spends much of the twenty-two-minute telecast trying to keep Earth in the frame, at first with no viewfinder. "I hope the next camera has a sight on it," he says. Mission commentator Haney stands at right.

Left to right: Trimble, Gilruth, and Low watch the image of the Western Hemisphere, which is improved after the astronauts have taped a polarizing filter to the TV camera's telephoto lens. Earth is 207,000 miles away.

Collins, Mattingly, Schmitt, Armstrong, and Aldrin at the Capcom console, with Dr. Hawkins at right. Flight director Milt Windler stands at upper left, with MSC photographer Charles Turner standing between consoles. Apollo 8 will soon enter lunar gravitational influence; the Moon is still about 40,000 miles ahead.

View of Earth broadcast by NBC-TV on December 23. "I hope that everyone enjoyed the picture that we're taking of themselves," says Anders.

Valerie Anders watches from the MOCR viewing room with James Elms, deputy associate administrator for manned spaceflight (*at left*). Lovell draws laughs when he suggests an approaching visitor from another planet might wonder whether to "land on the blue or brown part of the Earth."

Elms, Marilyn Lovell, and Kraft visit.

The Green team's flight director, Cliff Charlesworth, on December 23. An essential pencil sharpener is within easy reach at lower right.

Low, manager of the Apollo Spacecraft Program Office at MSC, would later serve as NASA deputy administrator in Washington, DC.

Journalists cover the flight in a temporary media work area in the Building 2 lobby on December 23.

NAR's Apollo Spacecraft program manager Dale Myers (*second from left*) and assistant NASA Apollo program manager Scott Simpkinson (*seated*) are in the Spacecraft Analysis Room, a support area behind the MOCR. Jim Tomberlin of the Apollo Spacecraft Program Office's Systems Engineering Division stands next to Simpkinson, with Owen Maynard, Tomberlin's chief, standing at right.

MSC recovery operations chief Donald Stullken, assistant mission director Chet Lee, and mission director William Schneider confer in the Recovery Operations Control Room on the right side of the MOCR.

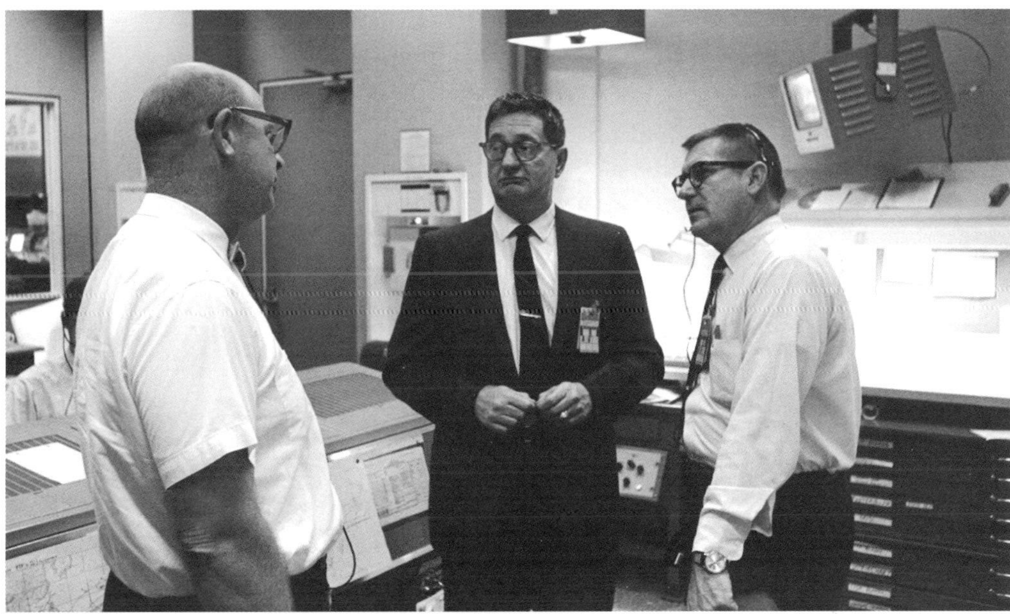

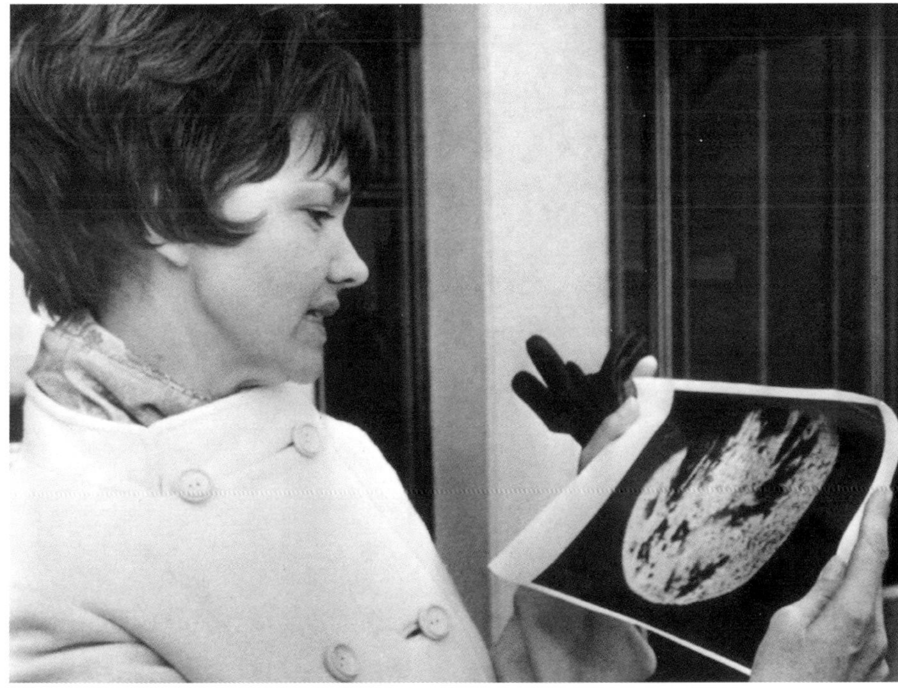

Marilyn Lovell outside the MCC, with a photo of Earth from the afternoon's TV transmission. *AP photo*

Thirteen-year-old Jay Lovell with a Rand McNally moon map at his home in Seabrook on December 23. *UPI photo*

Collins answers a question during the change-of-shift news conference in the Building 2 auditorium about 4:30 p.m. CST on December 23. *Left to right*: Haney, Charlesworth, flight dynamics officer Phil Shaffer, and Collins. *Photo by Jacques Tiziou*

Left to right: Haney, Charlesworth, Shaffer, and Collins. The shift had been problem-free, they report, with a focus on navigation and trajectory as the Moon, still about 8,750 miles away, grows closer. *Photo by Jacques Tiziou*

Shaffer explains some fine points of the spacecraft's location and velocity, with Charlesworth (*at left*). *Photo by Jacques Tiziou*

CHAPTER 11

December 24–26, 1968

A crescent moon rises above Building 30, the MCC, on the night of December 23 as Apollo 8 nears the Moon.

Right to left, seated: Black team controllers Tom Hanchett, Rod Loe, Sy Liebergot, Jack Kamman, Neil Hutchinson, Larry Canin, and Richard Thorson in in the MOCR. Standing behind Thorson are Buck Willoughby, Gary Cohen, and backup LM pilot Fred Haise, with Clint Burton behind Hanchett (*at right*).

At 2:53 a.m. CST on December 24, the astronauts get the "go for LOI [lunar orbit insertion]," and capcom Jerry Carr tells them they're "riding the best bird we can find." Advised by Carr at about 3:50 a.m. that LOS is one minute away, Anders says, "Thanks a lot, troops." Lovell adds, "We'll see you on the other side." The CSM then disappears behind the Moon, out of contact with Earth for about forty-five minutes, to conduct a four-minute burn of its SPS engine to break into an elliptical orbit.

"We've got it! Apollo 8 now in lunar orbit! There is a cheer in this room!" announces mission commentator John McLeish at 4:19 a.m. CST. Lovell radios, "Apollo 8, burn complete. Our orbit is 160.9 by 60.5 [nautical miles]." Carr replies, "Good to hear your voice." The flight will complete ten orbits at a velocity of about 3,000 mph. Each orbit takes about two hours.

Illustration of Apollo 8 in orbit

Aldrin (*right*), the backup CM pilot, goes over a chart with Schmitt and Armstrong, the backup commander (*left*).

The Black team's flight director, Glynn Lunney, at his console during the mission's third telecast, which starts at about 6:30 a.m. CST on December 24, during the second orbit, at almost seventy-two hours into the flight.

The crew shows the Moon for the first time, from an altitude of about 60 miles during the telecast, but the bright image has little detail at first. A lens filter helps, and soon some surface features become visible. "The color of the Moon looks, ah, a very whitish gray, like dirty beach sand, and with lots of footprints in it," says Anders.

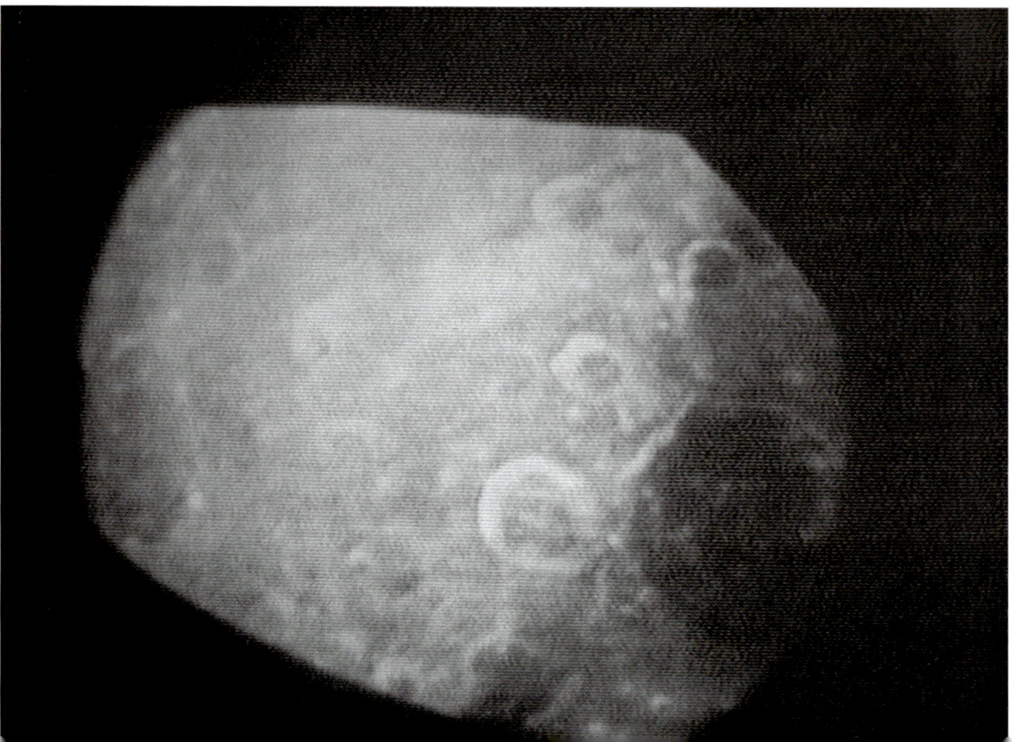

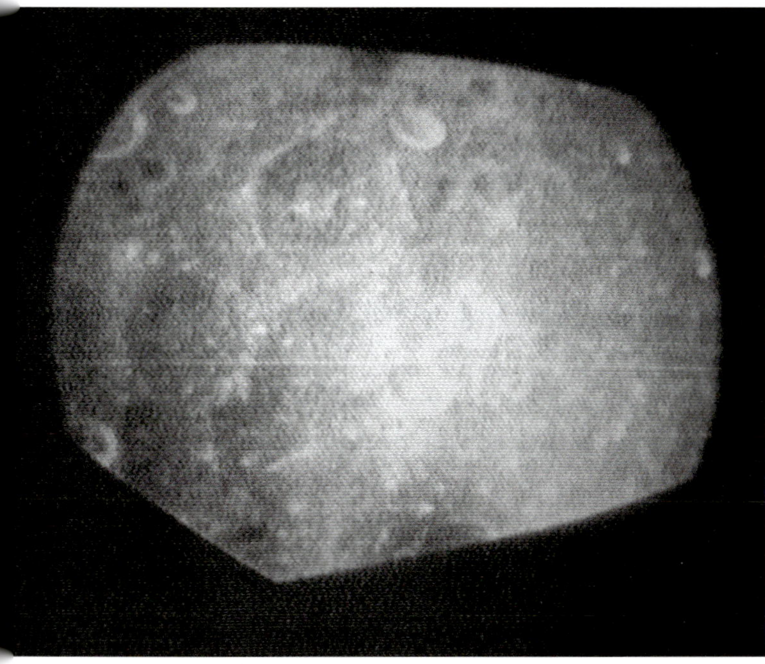

"These small impact craters have dark spots in the center," says Lovell, "where it appears that they buried in it and hit some new material down below and scattered a lot of fine white dust around them." This telecast lasts ten minutes and is the only TV received through the Madrid DSN station instead of Goldstone.

The Maroon team's flight director, Milt Windler, is present with the incoming Green team's flight director, Cliff Charlesworth (*left*). The spacecraft passes over several craters the crewmen have informally named for themselves, other astronauts, and NASA personnel.

Reporters in the Building 2 lobby serving as a media work area watch the TV transmission. *Photo by Jacques Tiziou*

A reporter composes his story. *Photo by Jacques Tiziou*

Just more than two hours later, reporters listen as Borman reports a second successful SPS burn behind the Moon to circularize Apollo 8's orbit at 8:42 p.m. CST. The 9.6-second burn was about twenty minutes earlier. The doors in the background lead to the auditorium for change-of-shift briefings. *Photo by Jacques Tiziou*

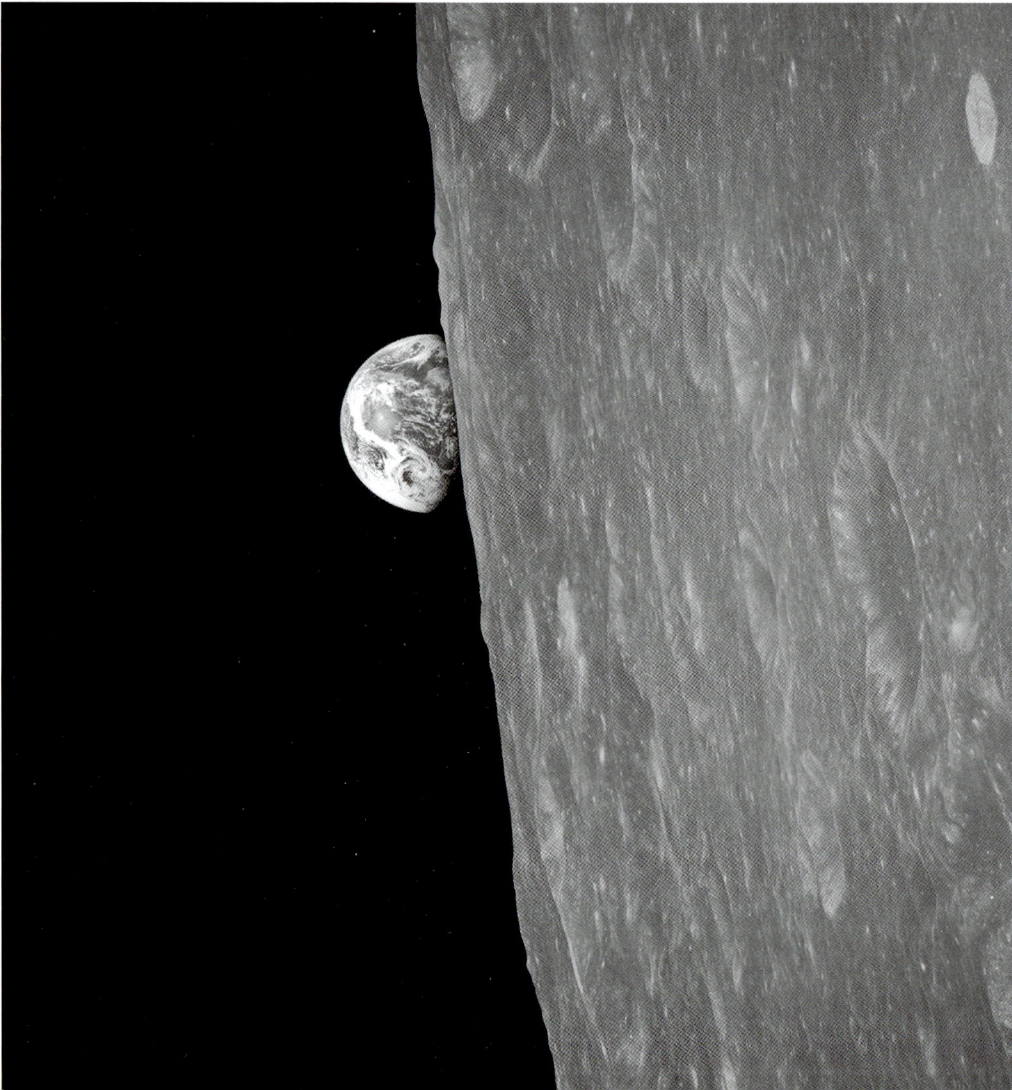

Apollo 8 is orbiting around the Moon's equator, and at the beginning of the fourth orbit on the far side, the astronauts see the Earth from behind the lunar limb. Anders takes the first such photo by an astronaut, looking across the middle of Pasteur Crater.

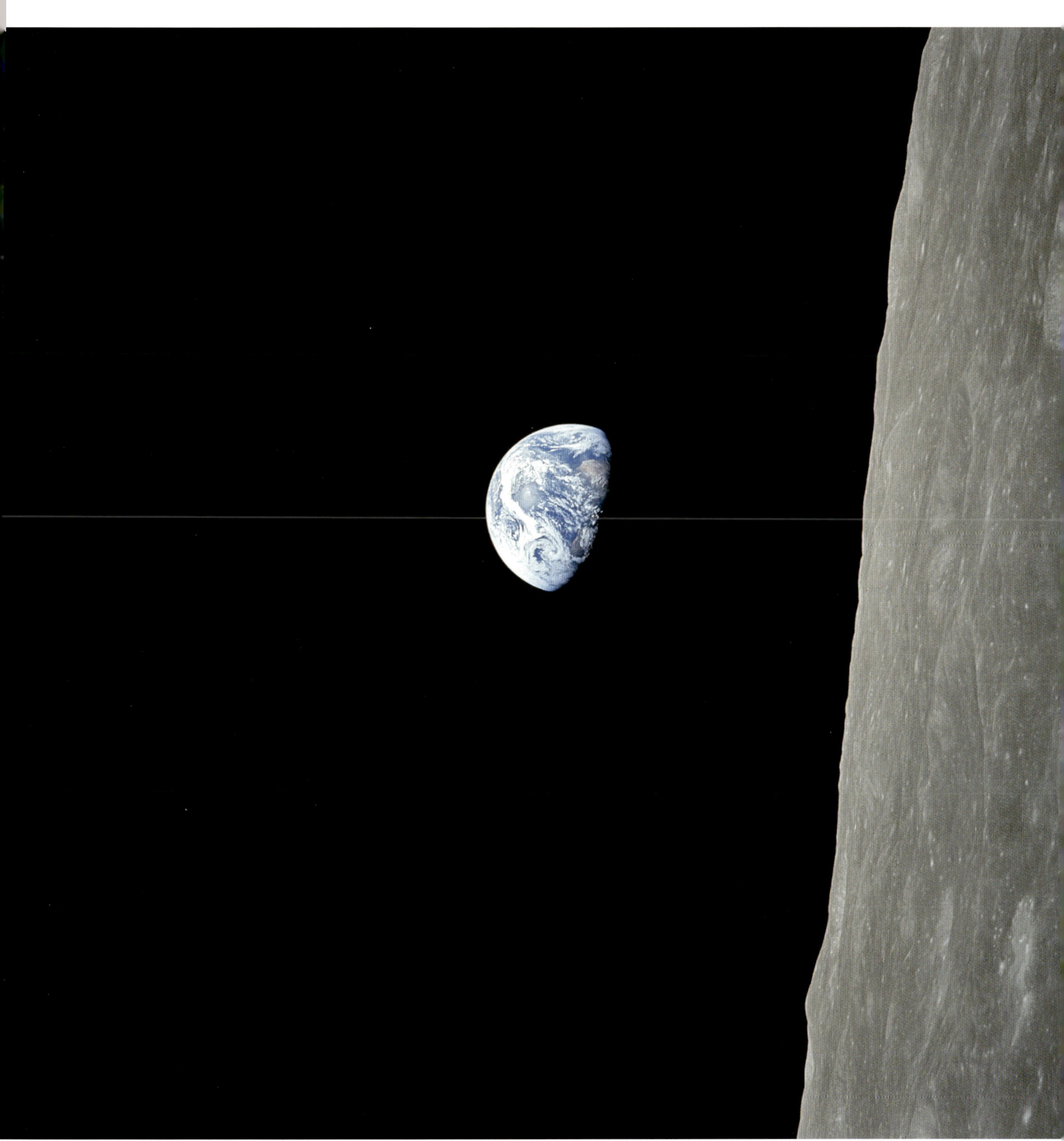

Moments later, Anders then snaps two color images, including the iconic photo eventually titled "Earthrise," at seventy-five hours and forty-eight minutes into the flight (10:38 a.m. CST), with Apollo 8 behind the Moon. "Oh, my God! Look at that picture over there!" he says. "Here's the Earth coming up. Wow, is that pretty!" He uses a 70 mm Hasselblad camera with a 250 mm telephoto lens and Ektachrome color reversal film.

The astronauts take 865 color and black-and-white photos of the lunar surface during their twenty hours in orbit, most on the far side, which has more sunlight. They are the first to directly see the far side of the Moon.

An astronaut observes the lunar surface in this cutaway NAR illustration of the CSM.

Extensive lunar-surface photography was a central goal of Apollo 8, especially of potential LM landing sites. The gear included two 70 mm motorized Hasselblad still cameras with a variety of lenses and filters, and one 16 mm Maurer data acquisition movie camera with several lenses. When mounted in a CM rendezvous window, it was used to shoot overlapping stereo frames at one frame per second along the lunar-orbit ground track, for terrain analysis. Some photography was hampered, however, because outgassing of fumes from the sealant around the windows caused fogging.

Crookes, a relatively young impact crater on the far side, is about 37 miles in diameter.

Crookes crater is in the foreground.

The Crookes X crater formation, west of Korolev Basin

Hirayama K crater is on the right, with Hirayama M at lower left on the far side.

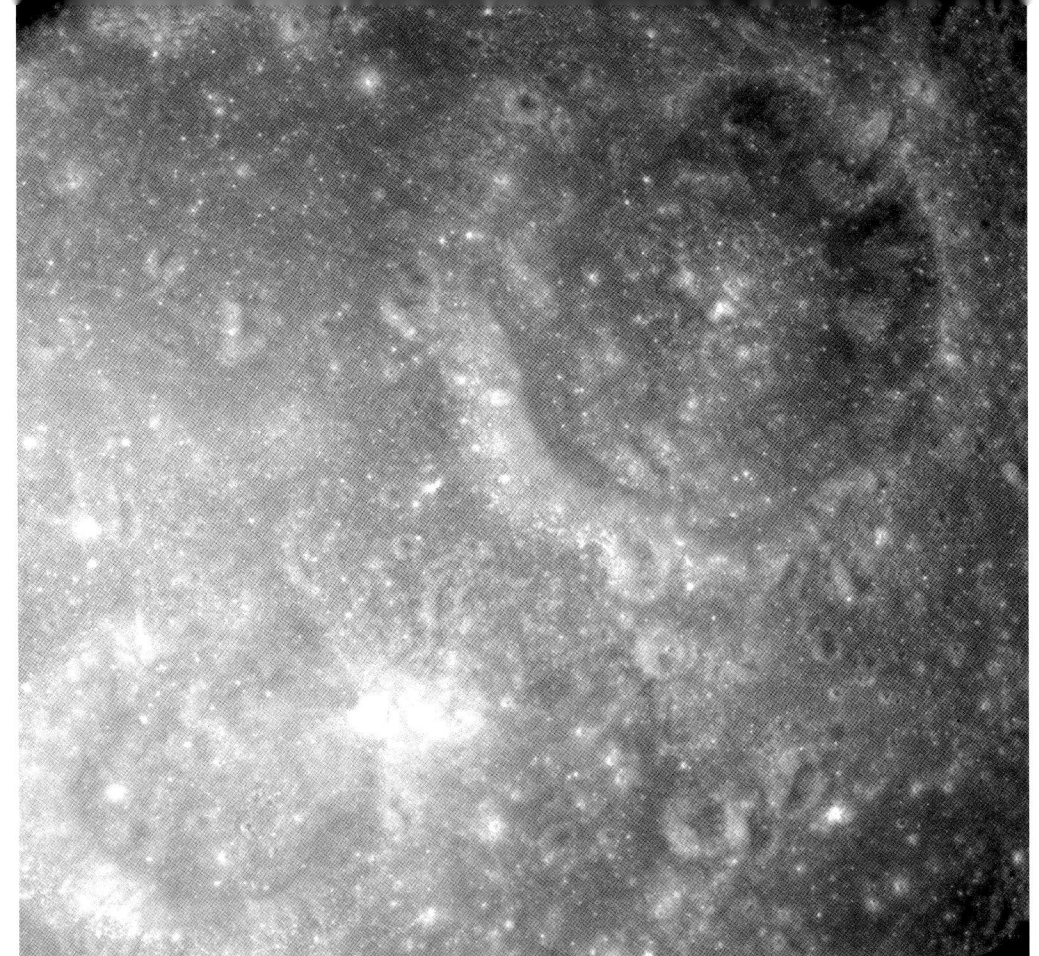

Looking west toward the Sea of Fertility

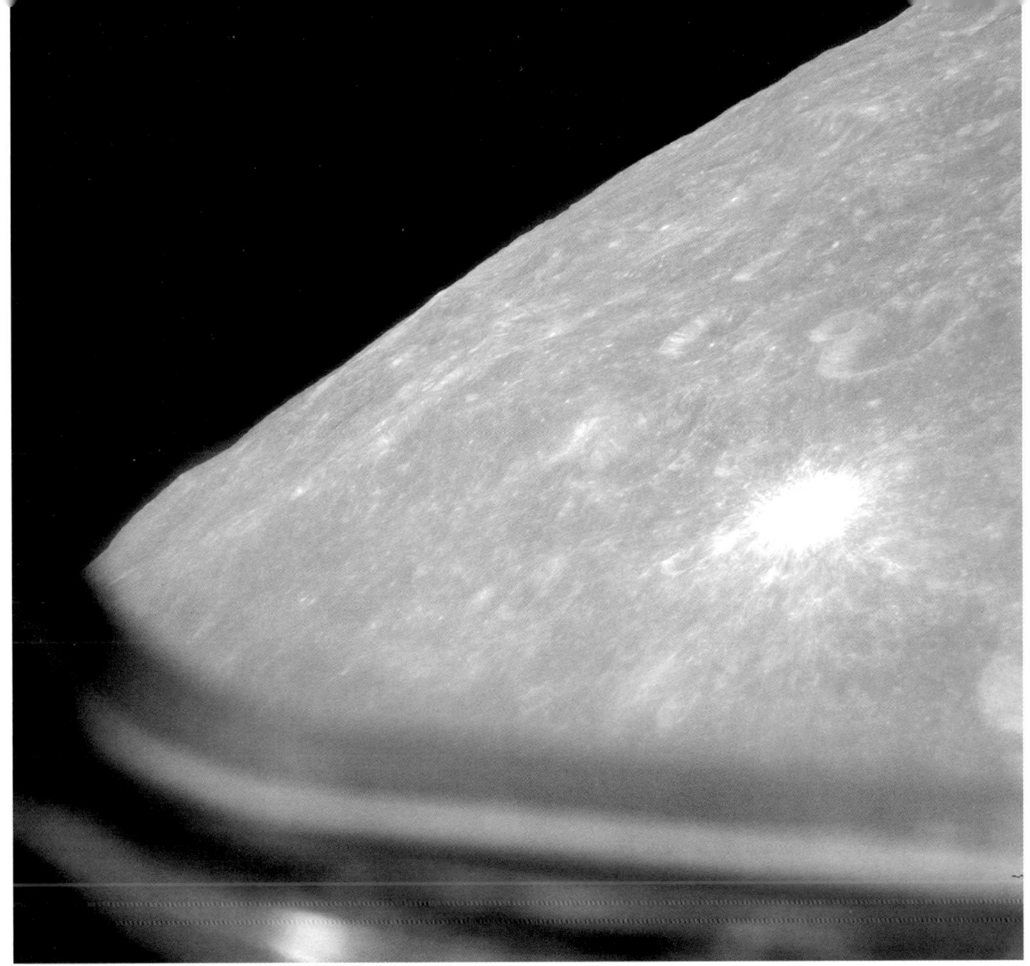

South of Tsiolkovsky crater

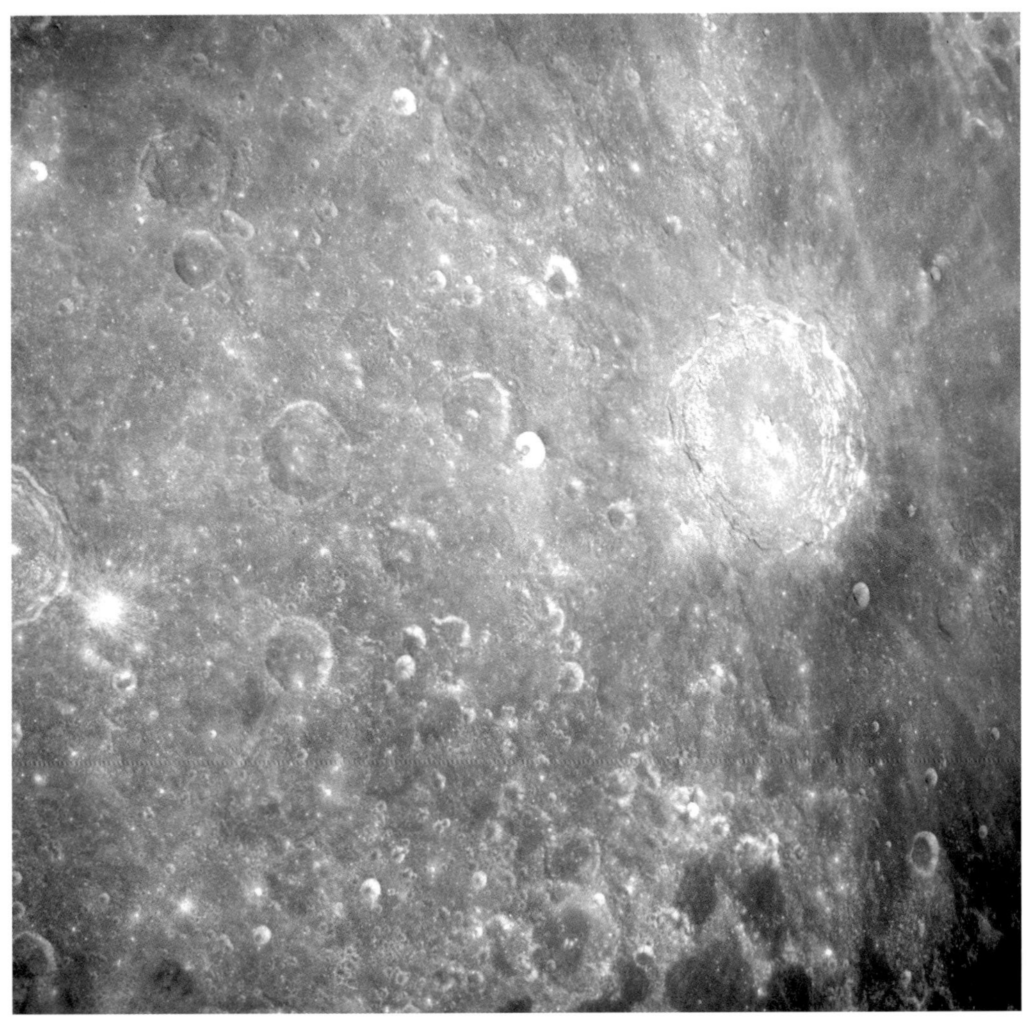

Langrenus crater is at center right.

Tsiolkovsky crater, photographed through a foggy window

Maskelyne F is the prominent feature left of center, during landmark tracking photography above the Sea of Tranquility.

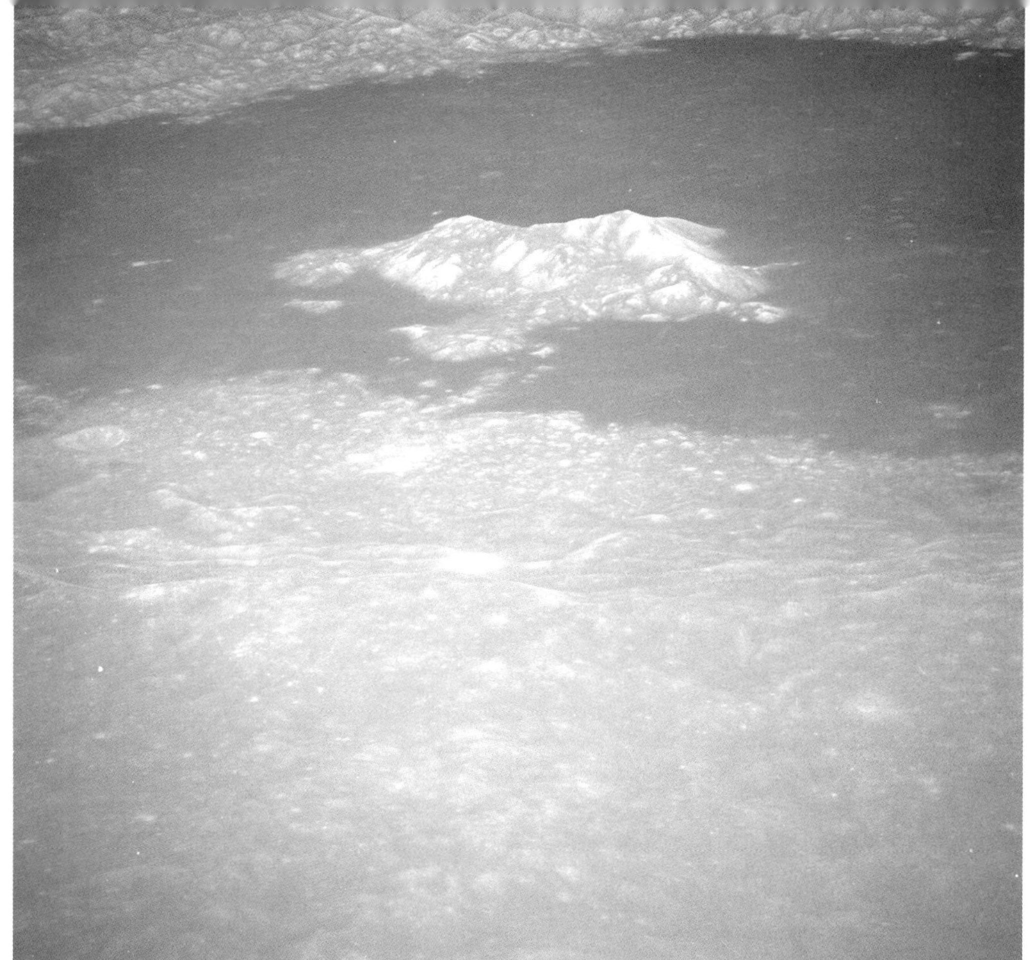

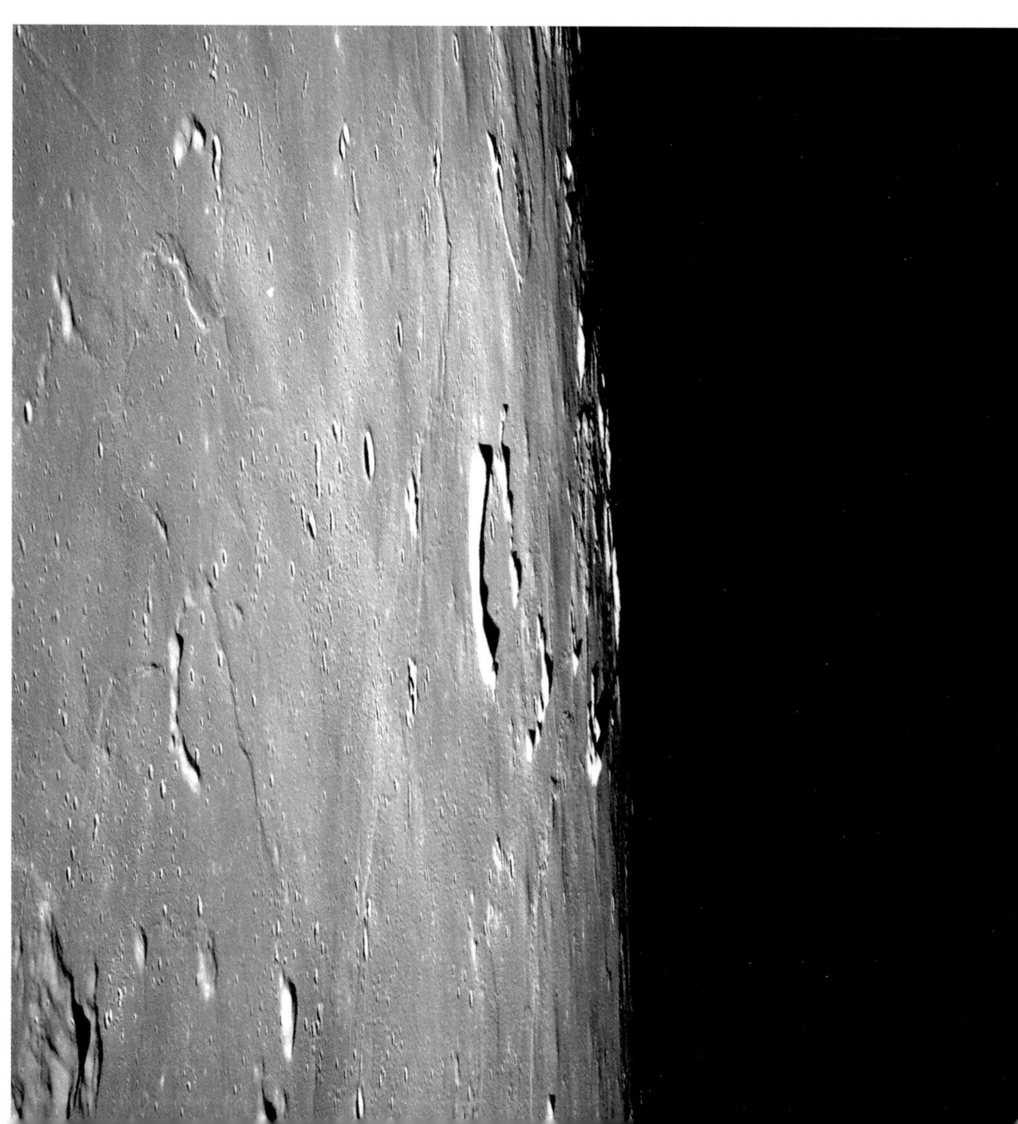

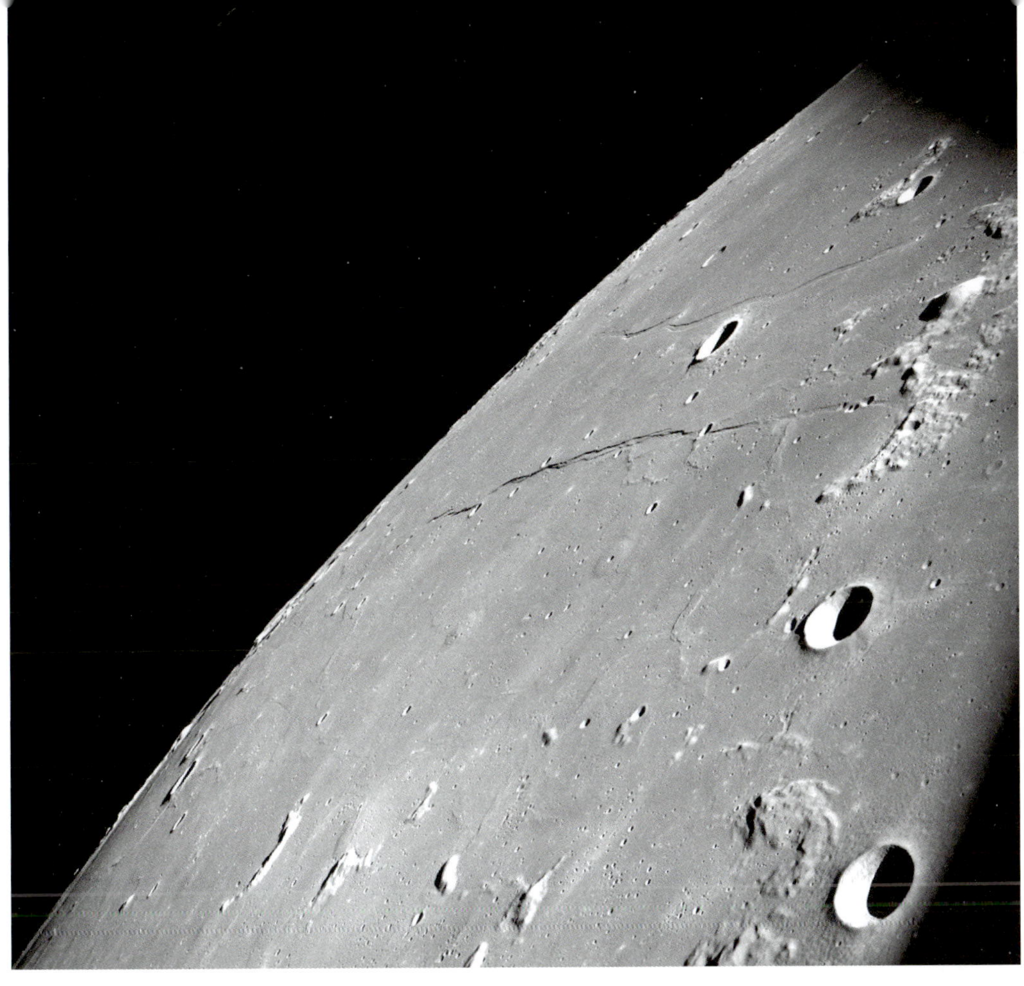

Cauchy crater is near the top, between two horizontal fractures in the Sea of Tranquility, looking west.

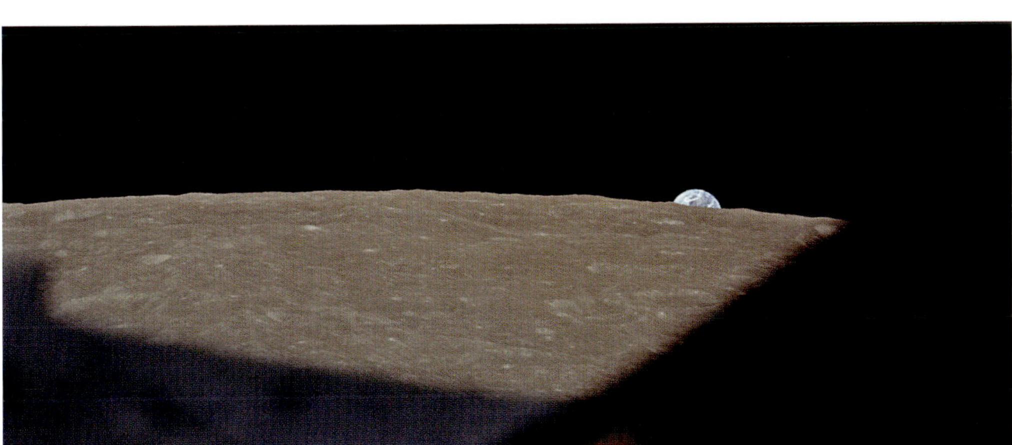

Earth appears above the lunar limb, looking west.

Same as previous image, but framed by a rendezvous window

This photo with a telephoto lens looks directly down on the far side. The area covered is approximately 40 square miles.

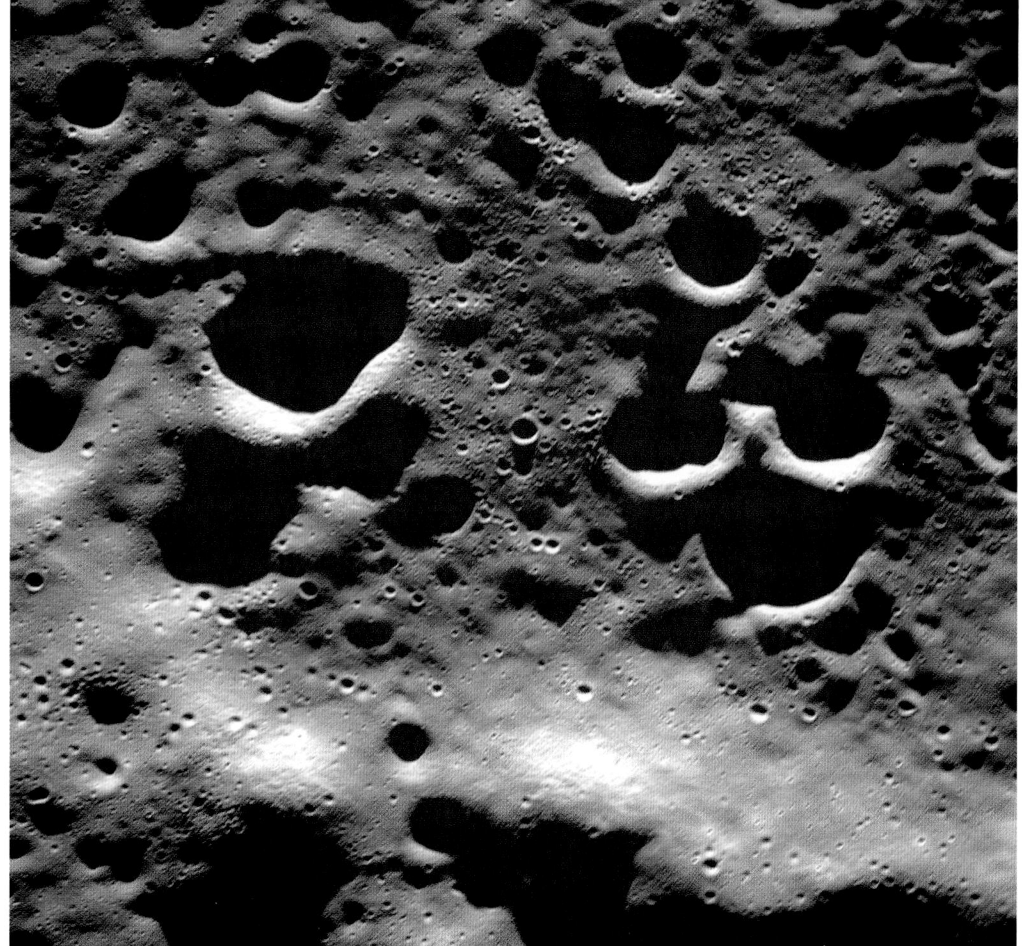

Amici T crater on the far side

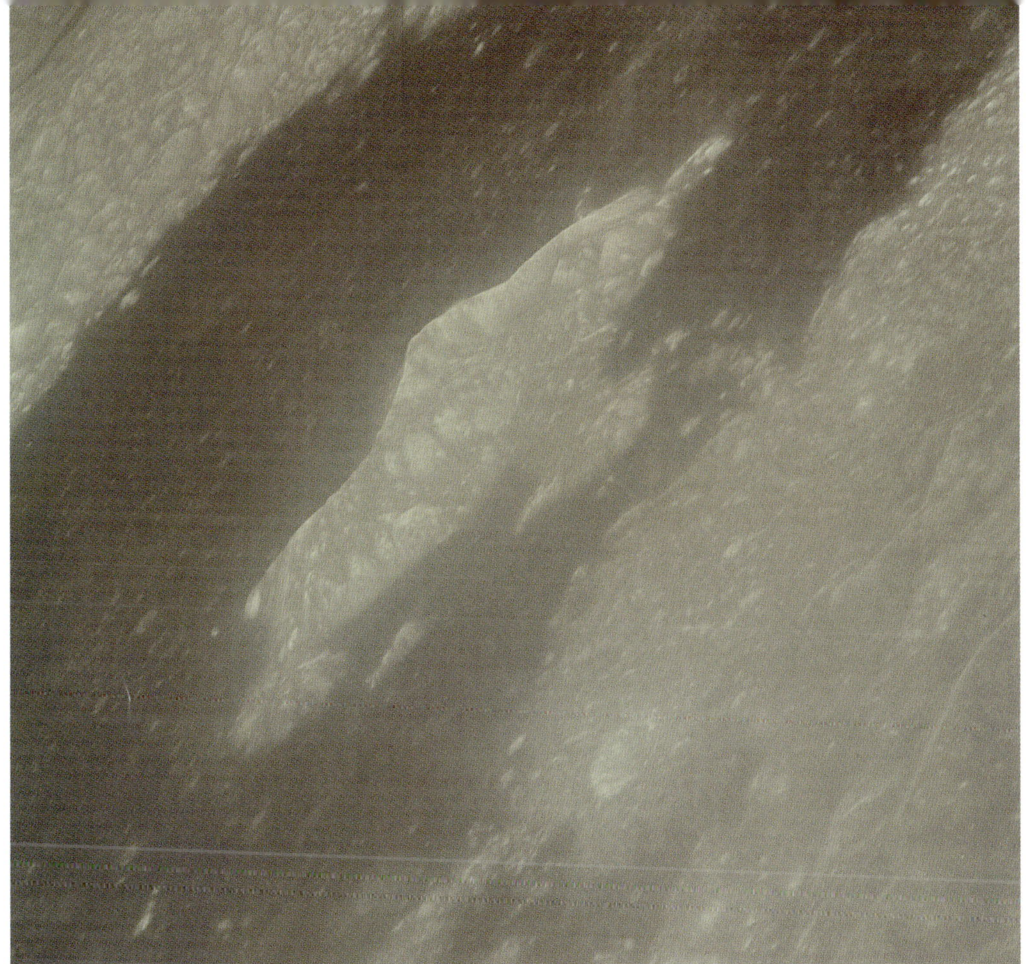

Tsiolkovsky crater

Langrenus crater

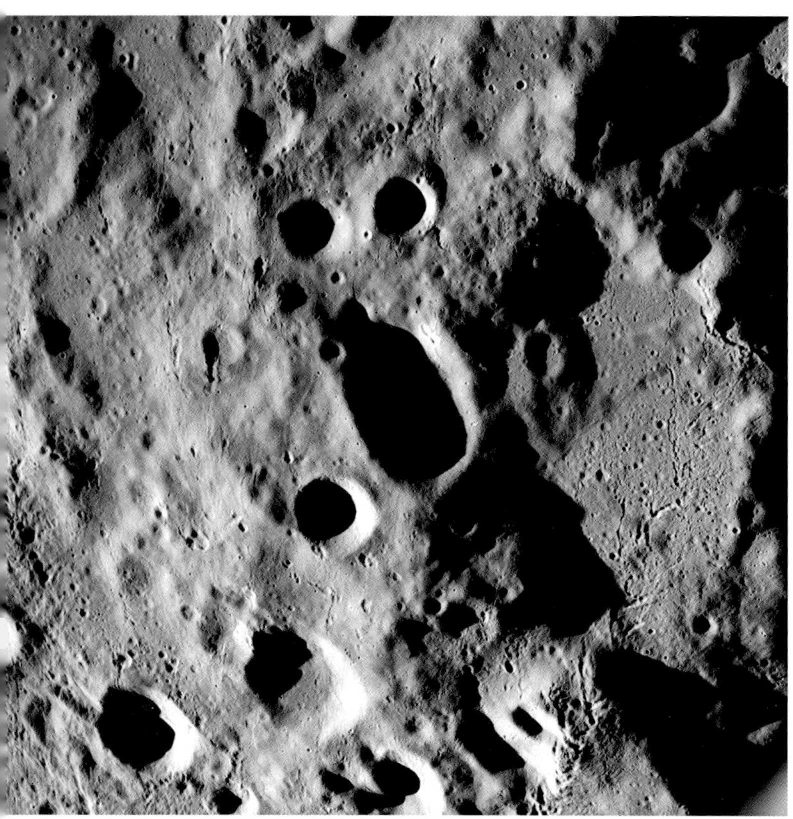

Far-side photo on the eighth orbit

Planté crater within Keeler crater, on the far side

Keeler S, west of Keeler

A far-side landscape south of Doppler crater

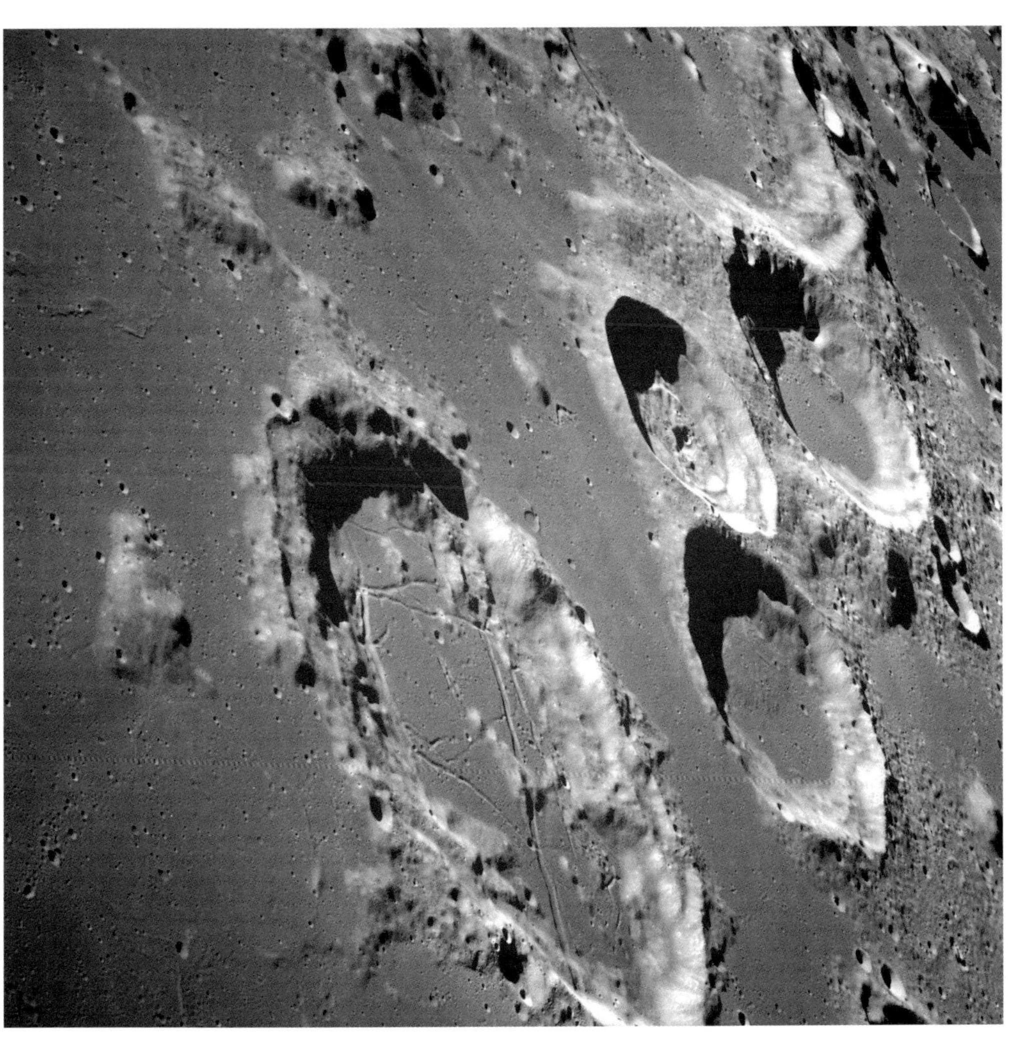

This view looks south in the Sea of Fertility. Goclenius, the large crater in the foreground, is about 45 miles in diameter.

The MCC on the evening of December 24

Armstrong in the MOCR during the fourth telecast on the mission's ninth orbit. The announcement that he would command Apollo 11 comes sixteen days later. "This is Apollo 8 coming to you live from the Moon," Borman begins.

View of the lunar surface during the telecast starting at 9:30 p.m. CST. The Moon, says Borman, is a "vast, lonely, forbidding expanse of nothing. It certainly would not appear to be a very inviting place to live or work."

Toward the end of the nearly twenty-six-minute telecast, as Apollo 8 crosses the Moon's terminator into darkness, the astronauts begin reading the first verses from the book of Genesis. They are moving west across the Sea of Tranquility, and their transmission is being watched by an estimated one billion people.

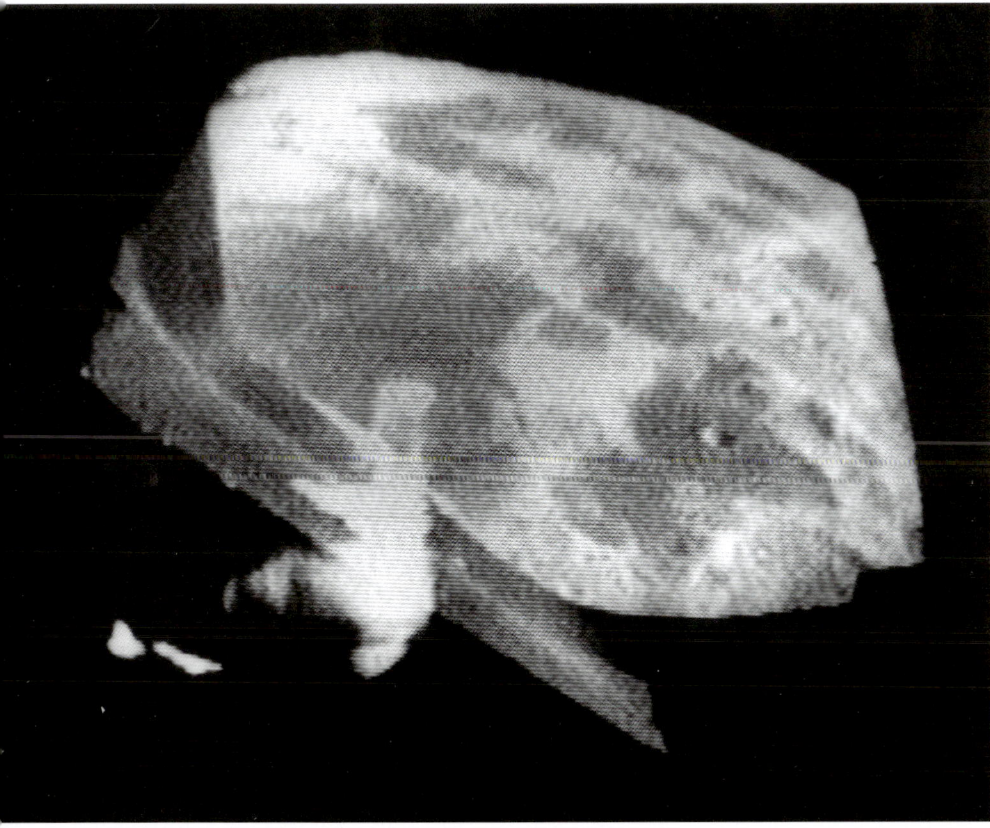

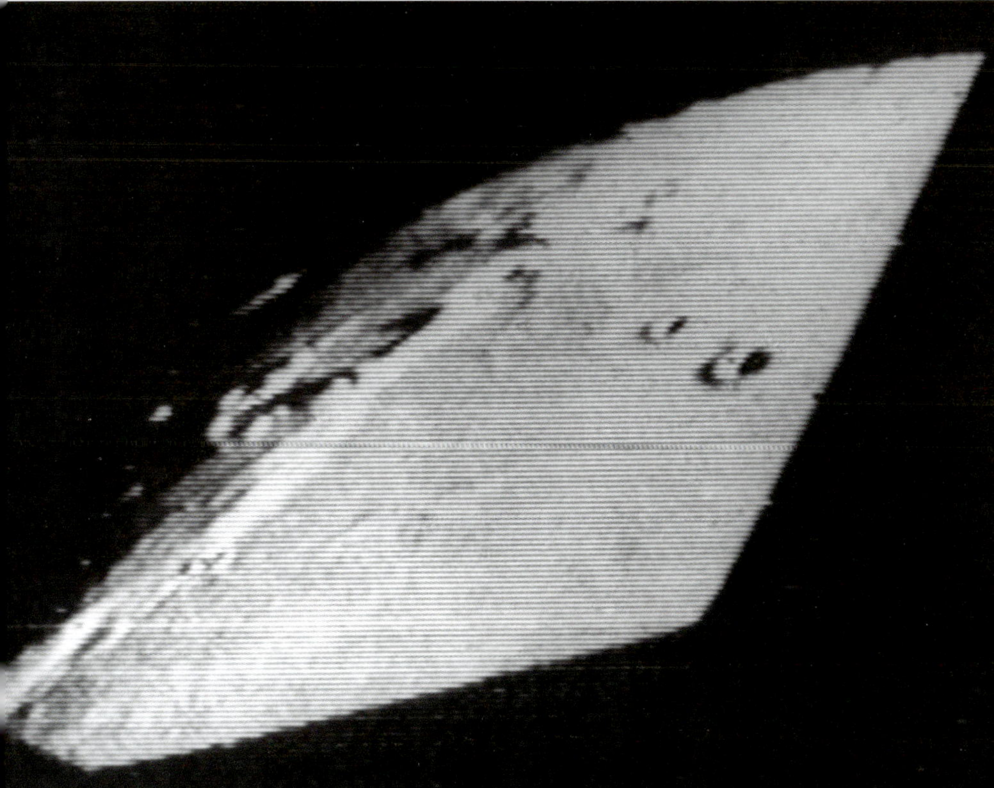

Anders: "We are now approaching lunar sunrise, and for all the people back on Earth, the crew of Apollo 8 has a message that we would like to send to you. In the beginning, God created the heaven and the earth. And the earth was without form and void, and darkness was upon the face of the deep. And the spirit of God moved upon the face of the waters, and God said, 'Let there be light.' And there was light. And God saw the light, that it was good, and God divided the light from the darkness."

Lovell: "And God called the light Day, and the darkness he called Night. And the evening and the morning were the first day. And God said, 'Let there be a firmament in the midst of the waters. And let it divide the waters from the waters.' And God made the firmament and divided the waters which were under the firmament from the waters which were above the firmament. And it was so. And God called the firmament Heaven. And the evening and the morning were the second day."

Borman: "And God said, 'Let the waters under the heavens be gathered together into one place. And let the dry land appear.' And it was so. And God called the dry land Earth. And the gathering together of the waters he called seas. And God saw that it was good. And from the crew of Apollo 8, we close with good night, good luck, a Merry Christmas, and God bless all of you—all of you on the good earth."

Just before 11:00 p.m. CST, capcom Mattingly radios the crew. "Okay, Apollo 8. We've reviewed all your systems. You have a go for TEI (trans-Earth injection)." Maroon team controllers gather at the unused LM consoles to monitor the flight. *Left to right*: Jim Hannigan, Rod Loe, Carrol Hopkins (USAF), Chuck Dietrich (*sitting on console*), and Jack Kamman (*hidden*). The display screen shows the status of the DSN stations.

About an hour later, as December 25 begins in Houston, an illustration shows Apollo 8's SPS engine firing behind the Moon for three minutes and eighteen seconds to leave lunar orbit. "Please be informed there is a Santa Claus," radios Lovell when they emerge. "We burned on time." Earth is 209,000 miles ahead.

Left to right: Schmitt, Aldrin, Mattingly, Slayton, and Armstrong in the MOCR early on December 25. Drs. Berry and Zieglschmid are to the right of Armstrong. Dr. Hawkins is behind Armstrong. "I just would like to wish you all a very Merry Christmas on behalf of everyone in the control center," says Slayton, "and I'm sure everyone around the world. None of us ever expect to have a better Christmas present than this one. Hope you get a good night's sleep from here on and enjoy your Christmas dinner tomorrow, and look forward to seeing you in Hawaii on the twenty-eighth."

The Moon recedes. Humboldt crater is on the left. Langrenus crater and the Sea of Fertility are on the right.

Jack Schmitt in the MOCR next reads a mission-oriented takeoff of "The Night Before Christmas" to the crew. An astronaut-geologist, he helped train Lovell for lunar observations.

A view across the Southern Sea. Jenner is the distinctive crater at right.

Tsiolkovsky crater dominates this view looking east.

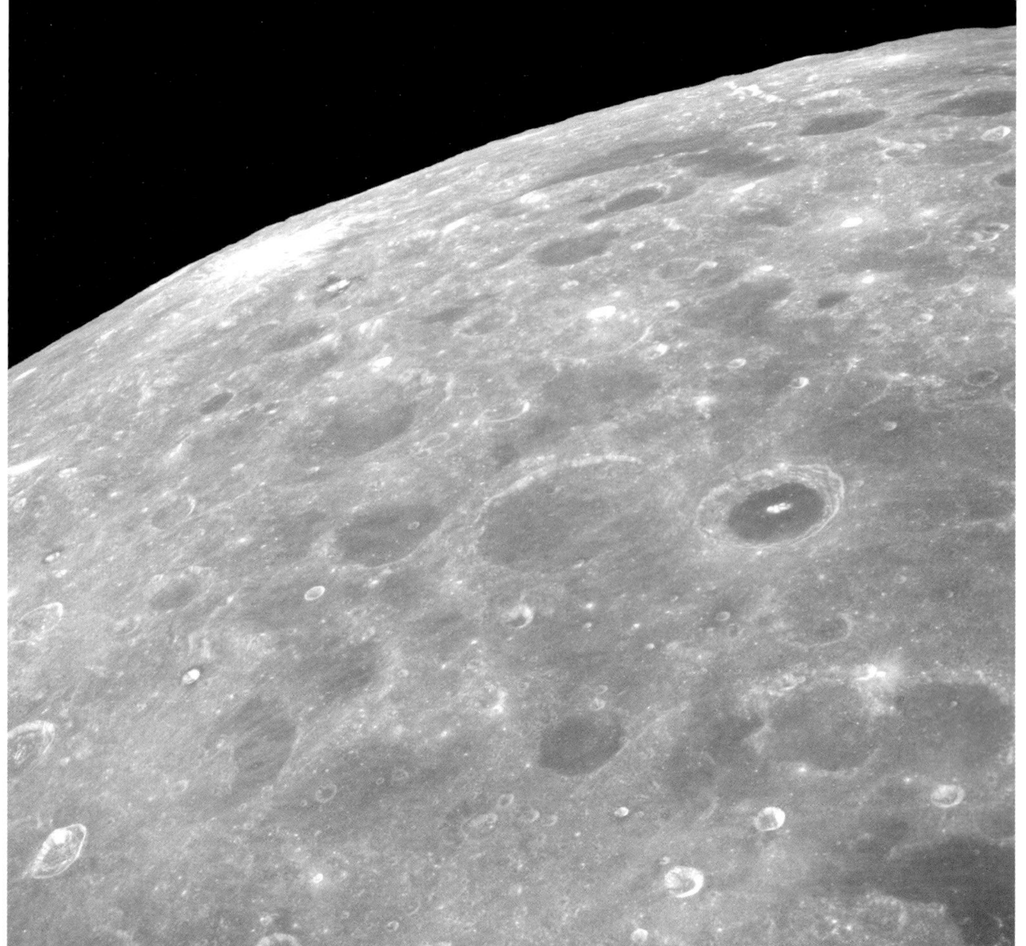

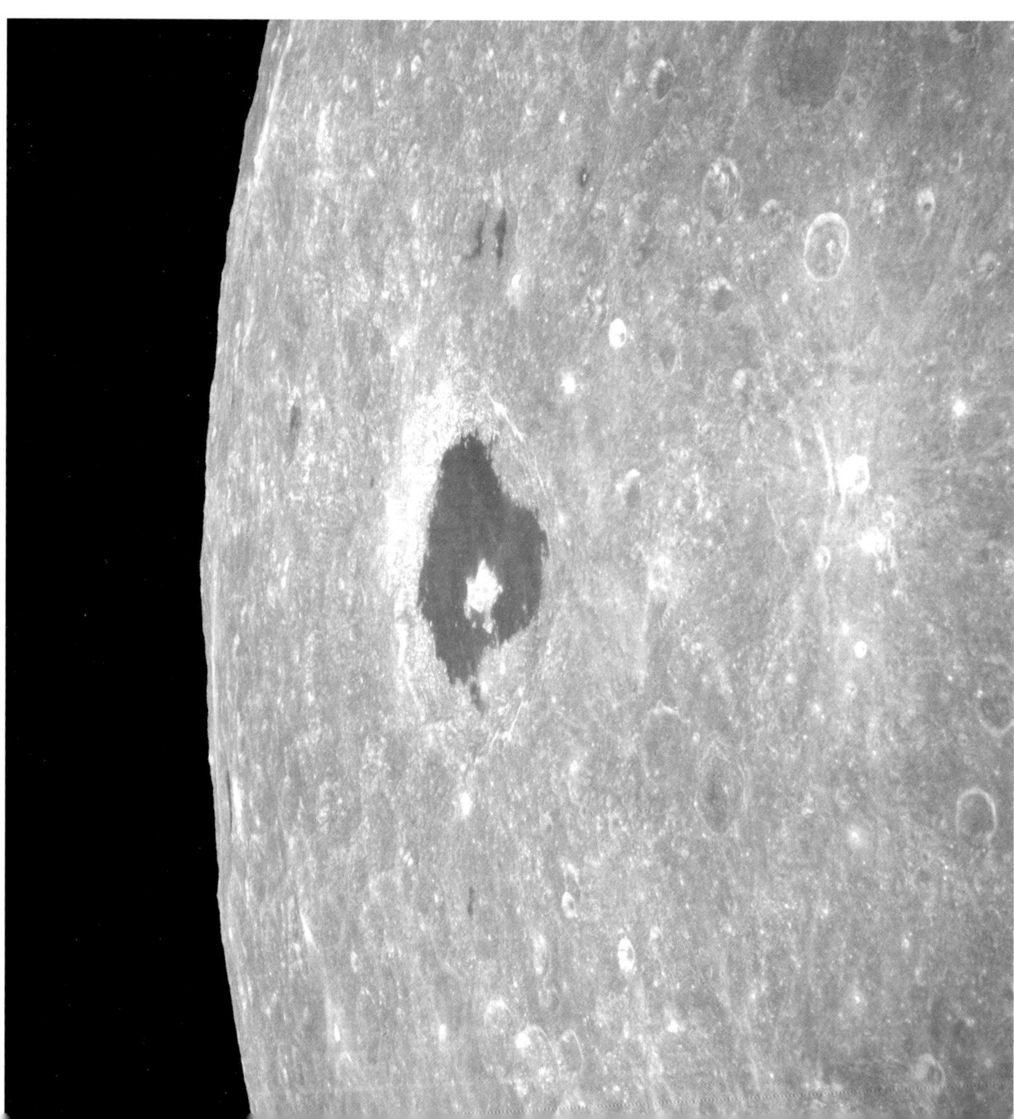

Susan Borman, outside St. Christopher Episcopal Church in League City on Christmas morning, brings audio tapes of the crew's Bible reading from the night before, to be played during the service. She says it's the nicest present she received. *AP photo*

Marilyn Lovell leaves for church in Seabrook that morning wearing a fur stole that had just been delivered as a Christmas gift from her husband. "It came from the man in the Moon," she tells reporters. *UPI photo*

Valerie Anders and sons Glen (age ten), Gregory (six), and Eric (four) arrive at the Ellington AFB chapel for Christmas services, about a thirty-minute drive west of Seabrook. *AP photo*

During the mission's fifth telecast, about 3:10 p.m. CST that day (104 hours into the mission), Borman provides a tour of the cabin. Lovell extends his legs in the foreground as part of daily exercise. Anders will demonstrate food preparation.

In the lower equipment bay, Lovell explains the optics for the guidance system during the ten-minute telecast.

Left to right: Mission commentator John McLeish (*hidden*), flight director Glynn Lunney, capcom Jerry Carr, flight dynamics officer Ed Pavelka, and an unidentified official at a briefing after the Black team's shift. *Photo by Jacques Tiziou*

Left to right: McLeish, Lunney, Carr, Pavelka, and an unidentified official. *Photo by Jacques Tiziou*

Lunney at the briefing. *Photo by Jacques Tiziou*

Apollo 8 leaves the Moon behind.

A nearly full moon comes into view. The Sea of Crisis, the circular, dark-colored area near the center, is near the eastern edge. The large, irregular seas are Tranquility and Fertility. Tsiolkovsky crater is near the limb at the lower right.

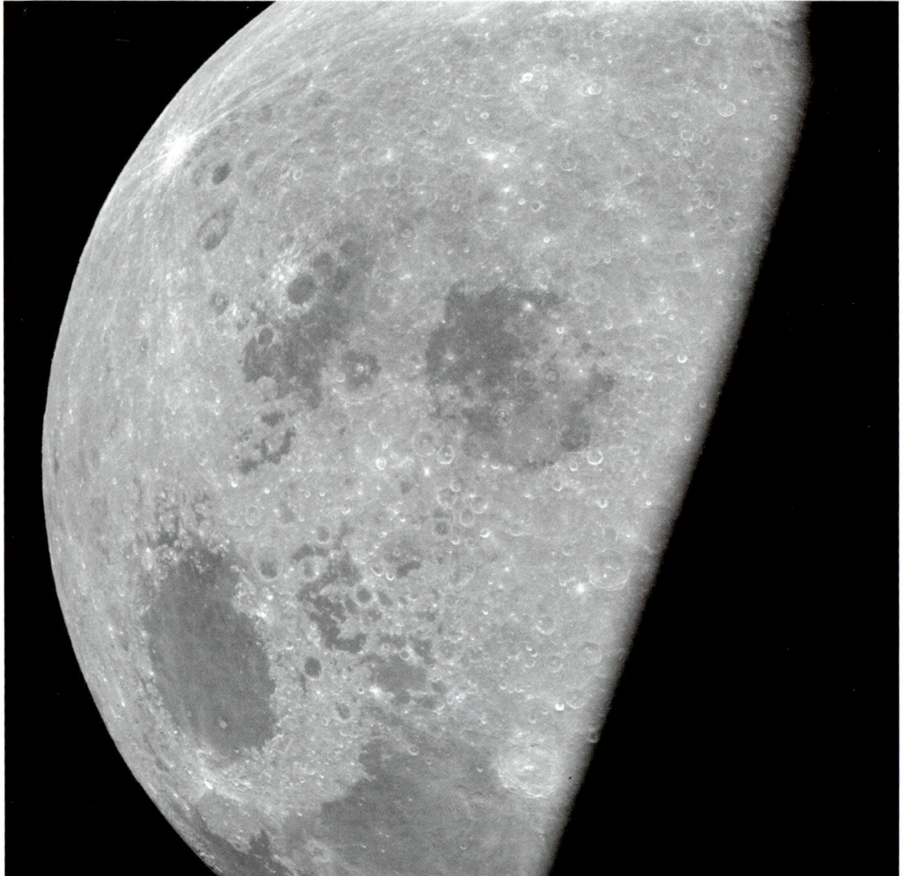

These two photos are among thirty-three accidentally taken on 70 mm color film a few seconds apart, intended to be a colorimetric experiment to better analyze color differences in the Moon as observed from Earth, using a red or blue filter with black-and-white film. The color film accidentally used, however, produced these images, which were not useful. The crew did then shoot some red-and-blue pairs in black and white but again used the wrong (dim-light) film, which was overexposed during processing.

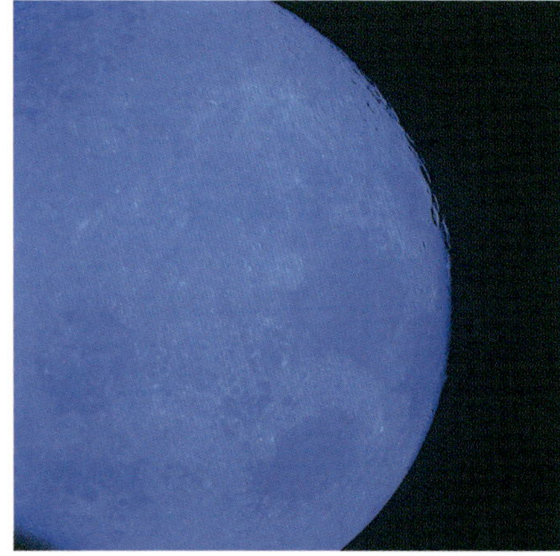

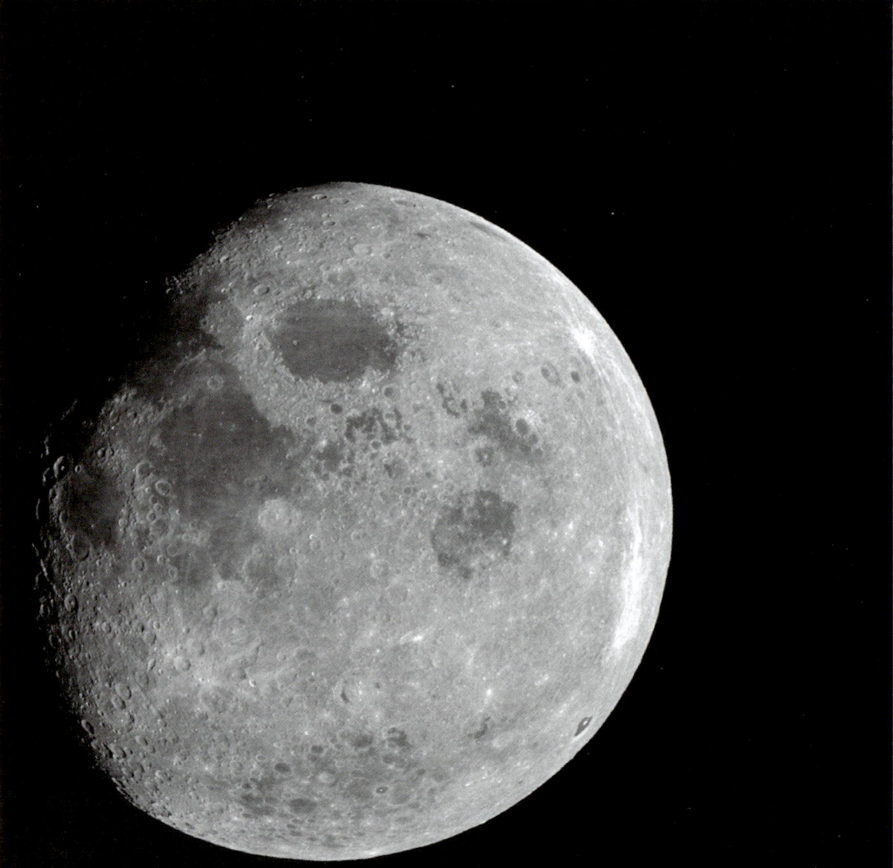

Africa is visible as the Earth grows closer. “We’re happy to report the earth is getting larger,” radios Borman.

Anders holds a Hasselblad camera (16 mm still frame).

Lovell pulls himself into the lower equipment bay; Anders is in the background (16 mm still frame).

Borman in his couch (16 mm still frame)

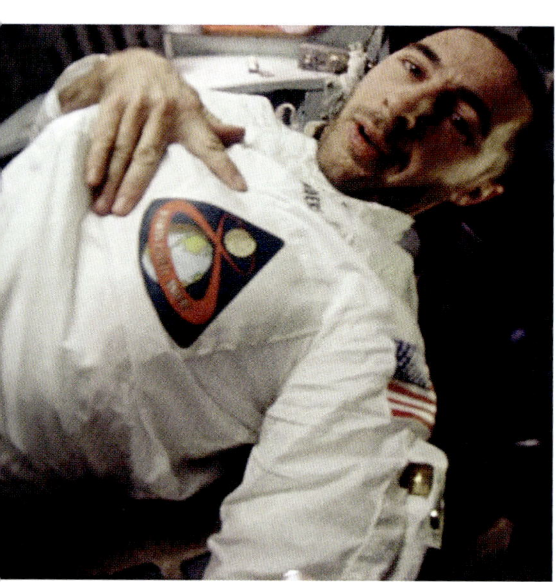

Anders displays the mission emblem (16 mm still frame).

Apollo 8 nears Earth in this illustration.

Lunney, the Black team's flight director, and Kraft monitor the mission's sixth and final telecast about 2:50 a.m. CST on December 26, 128 hours into the mission. Schmitt, Aldrin, and Carr huddle at the Capcom console at center.

Low, Kraft, and Lunney watch the twenty-minute transmission. "At the tip of South America, there's a great swirl of clouds down there," says Lovell. "It looks like a great storm." Two boxes of cigars at lower right are ready for splashdown the next day.

Lovell's family watches from the MOCR viewing room. *Front to back*: Jay Lovell, Barbara Lovell, and Marilyn Lovell, and Alan Shepard with daughter Laura Shepard.

View of Earth at 191,000 miles. "As I look down on the earth here from so far out in space," says Anders, "I think I must have the feeling that the travelers in the old sailing ships used to have: Going on a very long voyage away from home, and now we're headed back, and I have that feeling of being proud of the trip, but still happy to be going back home and back to our home port. And that's what you're seeing right here."

"We've enjoyed the television shows," says Borman, "and we'd like you to stay tuned in, in the future, because there'll be flights and rendezvous in Earth orbit, and then, of course, there'll be television from the lunar surface itself in the not-too-far-distant future. So, until then, I guess this is the Apollo 8 crew signing off, and we'll see you back on the good Earth very soon."

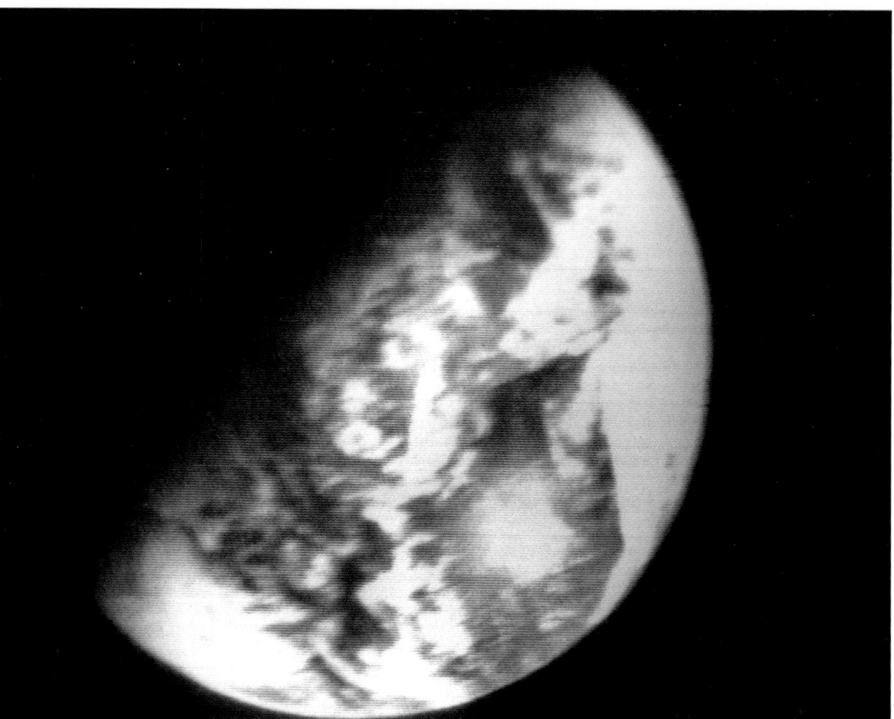

Susan Borman, Valerie Anders, and Marilyn Lovell attend a party hosted by Mary Haise, wife of backup LM pilot Fred Haise. *UPI photo*

A half moon is above the marquee at the Ramada Inn on NASA Road I west of MSC; the G is missing in Gemini Club. *Photo by Jacques Tiziou*

Marquee at the Nassau Bay Shopping Village across NASA Road I from MSC. *Photo by Jacques Tiziou*

CHAPTER 12

December 27, 1968

The aircraft carrier USS *Yorktown* (CVS-10), with its crew of 1,650, is on station about 1,000 miles southwest of Hawaii in the Pacific the day before splashdown. "This is the most isolated manned recovery we've had yet," says MSC Recovery Team leader John Stonesifer. Just before leaving San Diego on December 5 as the Apollo 8 prime recovery ship, *Yorktown* had been used for three days as a stand-in for the lead Japanese carrier, *Akagi*, to reenact the 1942 Pearl Harbor attack for the 1970 feature film *Tora! Tora! Tora!*

Chaplain Dean Veltman leads a joint Catholic-Protestant prayer service on the flight deck on Christmas morning.

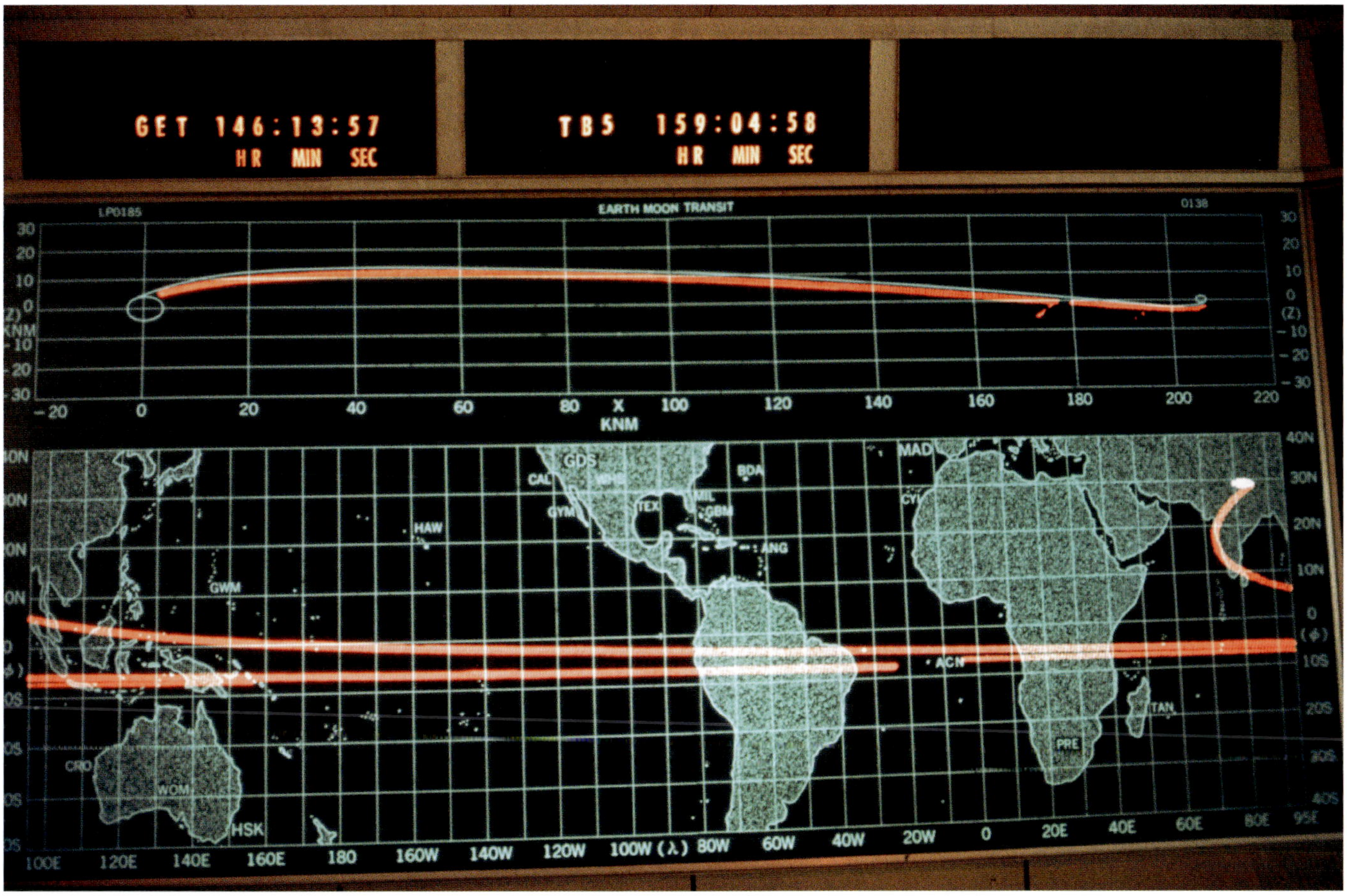

The MOCR display screen has changed from the lunar mission to an Earth map shortly before the crew is given a go for CM/SM separation, about thirty minutes before splashdown on December 27.

Low (*right*) chats with flight director Lunney.

Left to right: Black team flight controllers Tom Hanchett, Charlie Dumis, Sy Liebergot, Clint Burton, and Joe DeAtkine at the CM EECOM (Electrical, Environmental, and Communications) console

Left to right: Phillips, Kraft, Low, Gilruth, and Trimble monitor the reentry.

Apollo 8 flight directors Charlesworth (*left*) and Windler (*right*) flank Lunney.

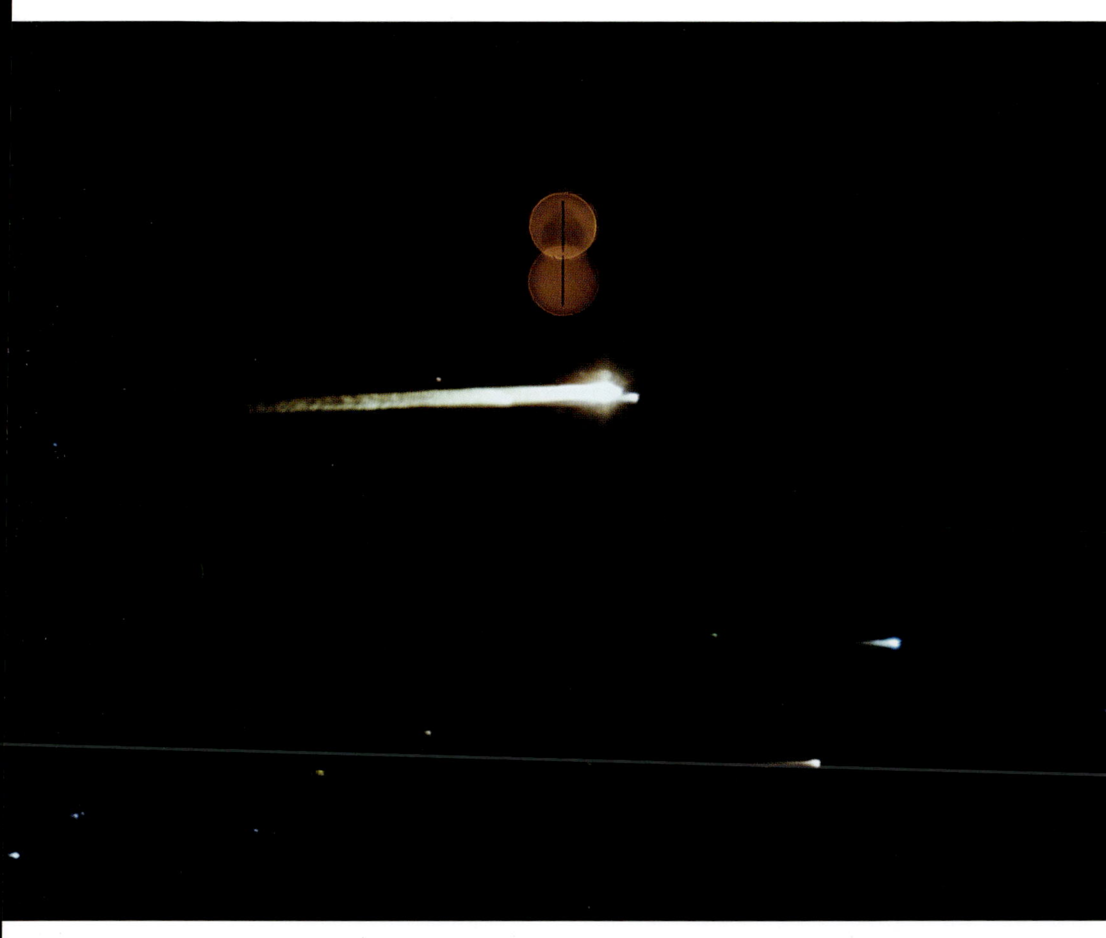

The fiery reentry is photographed by a 70 mm Airborne Lightweight Optical Tracking System camera, mounted in a pod on a cargo door of a USAF EC-135N aircraft.

Lunney waits during a three-minute communications blackout caused by the ionized, heated air surrounding the CM.

Phillips and Low appear confident.

Yorktown's flight deck crew prepares for the astronauts' arrival as the carrier heads toward the CM. The white dome at right protects a Western Union uplink to relay live color TV via NASA's ATS-4 satellite.

Primary Flight Control on the superstructure is busy shortly after sunrise.

Apollo 8 began to reenter the atmosphere at 4:15 a.m. Hawaii Standard Time (HST) at an altitude of 76 miles. The CM had achieved a record speed of 24,171 mph, and the heat shield reached 4,740 degrees Fahrenheit. Although Apollo 8 was the first manned splashdown in darkness—at 4:51 a.m. HST—it landed so close to its target that one recovery helicopter could see its parachutes, and its flashing beacon was visible from USS *Yorktown*, which was less than 3 miles away. After splashdown, the choppers could see the CM's cabin lights as they circled overhead for about forty-five minutes until dawn, when swimmers were deployed.

Three members of Underwater Demolition Team 12 attach a flotation collar and hold two life rafts dropped by a chopper. The Navy Sikorsky SH-3D Sea King helicopters are from Helicopter Anti-Submarine Squadron (HS)-4. "Hello there, Houston; how you doing?" Lovell asks.

Borman joins Lovell in a life raft about 6:05 a.m. HST. Asked earlier by the Helo 66 pilot, Navy commander Don Jones, what the Moon is made of, Borman replies, "It's not made of green cheese at all . . . it's made out of American cheese." The three uprighting bags are inflated; the CM was briefly inverted after splashdown.

Sonar Technician (STG) Bob Coggin (*left*) and Lt. Richard Flanigan assist Anders, the last to egress under cloudy skies with showers in the area.

Helo 67 hovers near Apollo 8, with all three astronauts in the raft.

Borman is hoisted to the helicopter in a Billy Pugh net. "I think this is a real good piece of gear," he says as he looks down at a shark.

Lovell waits in the raft.

Lovell is hoisted next.

Anders is raised to the chopper for the six-minute flight to *Yorktown*.

Helo 66 prepares to land on the deck of *Yorktown* at 7:20 a.m. HST, one hour and twenty-eight minutes after splashdown. The arrival is carried live by the three US TV networks, with ABC serving as pool producer.

Borman, Lovell, and Anders wave to the hundreds of crewmen, who applaud, cheer, and snap photos.

Photographers capture the astronauts after they emerge.

Borman waves to sailors on upper levels.

The crewmen steady themselves after more than six days in weightlessness. Borman, who'd requested an electric razor be aboard, had shaved en route.

Kraft and Low shake hands and light up cigars as flight controllers from all three shifts fill the MOCR. A large American flag is unfurled, and "The Star-Spangled Banner" is played over a communications loop. "I have never seen a degree of this emotional outpouring in any previous mission, including Alan Shepard's," says mission commentator Paul Haney. "I guess one of the big differences there between that one and this one is Alan is here standing right in the middle of this one, puffing on a long black cigar (*far left*). I've seen rallies in locker rooms after championship games, happy politicians after elections, but never—none of them do justice to the spirit pervading this room." It's almost noon CST in Houston.

Lovell, Borman, and Anders head toward the microphone along a red carpet.

Capt. John Fifield greets the astronauts. "On behalf of the entire *Yorktown* crew, a most hearty welcome aboard, and congratulations on a tremendously successful flight," he says.

Borman addresses the sailors, with a US Marine Corps honor guard behind the crew. “We’re just very happy to be here,” he says, “and we appreciate all your efforts.”

"I know you had to stay out here over Christmas, and that made it tough," he continues. "Jim and I always seem to fly in December. We made it home before Christmas in '65 [on Gemini VII]."

"But we can't tell you how much we really appreciate you being here and how proud it is for us to participate in this event, because thousands of people made this possible and I guess we're all just part of the group. Thank you very much," Borman concludes.

Gilruth, Trimble, Kraft, and Low celebrate in MOCR. In rear are an unidentified person, mission director George Hage, Phillips, and MSC director of engineering Max Faget.

Borman, Anders, and Lovell don *Yorktown* "Fighting Lady" caps from Capt. Fifield. Dr. Clarence Jernigan, flight surgeon, is between Borman and Lovell, with NASA public affairs officer Ben James at far right.

Capt. Fifield (*right*) escorts the astronauts, with the honor guard following.

Lovell, Anders, and Borman head to the elevator to the hangar deck with Dr. Jernigan (*at right*).

The astronauts interrupt four hours of medical tests in sick bay for breakfast. "We'll have steak and eggs, the same as we had before we left," one had radioed when asked about the menu by *Yorktown* during the wait for daylight.

Lovell reads a message of congratulations as Dr. Jernigan looks over his shoulder.

The crewmen take a second break to listen to a message from President Lyndon Johnson he'd recorded in the Fish Room at the White House (Johnson recorded his comments because of the uncertainty of the astronauts' availability). "You have made us very proud to be alive at this particular moment in history," he says. "You have made us feel akin to those Europeans nearly five centuries ago who heard stories of the New World for the first time. There is just no other comparison that we can make that is equal to what you have done or to what we feel."

Borman listens to Johnson's message, which discloses that the hotline teletype was used to keep Moscow posted on the flight's progress. "The Soviets were very solicitous about the welfare of you astronauts and expressed great interest in the success of your flight," he says.

Lovell listens. "If I could have exchanged thoughts with you, I was going to ask you whether it felt better coming down or going up, and to have you tell me some of your experiences, because you have seen what man has really never seen before. You have taken all of us all over the world into a new era."

Anders listens to the president's recording. "We rejoice that you are well, and we send you congratulations from all of your fellow countrymen and from all peace-loving people in the world. Well done," he concludes. Johnson had talked with their wives on a conference call about an hour earlier.

Lt. Flanigan (*top*) and STG Coggin wait to attach the scorched CM to *Yorktown*'s crane after the carrier arrived nearby at 7:13 a.m. HST.

The swimmers prepare to attach the recovery hook to the CM, so personnel from the ship's Weapons Department can lift it with cranes.

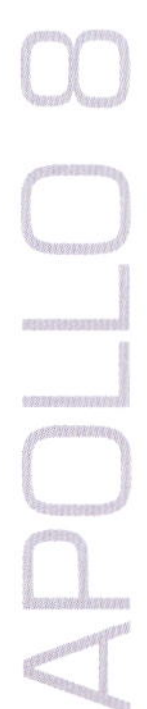

Lt. Flanigan (*top*) and STG Coggin make final preparations for the CM to be raised.

Lt. Flanigan signals to *Yorktown* personnel as STG Coggin assists.

The CM is hoisted by its recovery loop.

Two hours and twenty-eight minutes after splashdown, the CM is brought onto Elevator 3.

The CM is on its dolly to be towed into the hangar deck.

NASA recovery personnel make initial inspections of the CM.

One pair of RCS roll thrusters on the CM show the effects of reentry.

A view of the aft heat shield shows one of the three attach points for the SM.

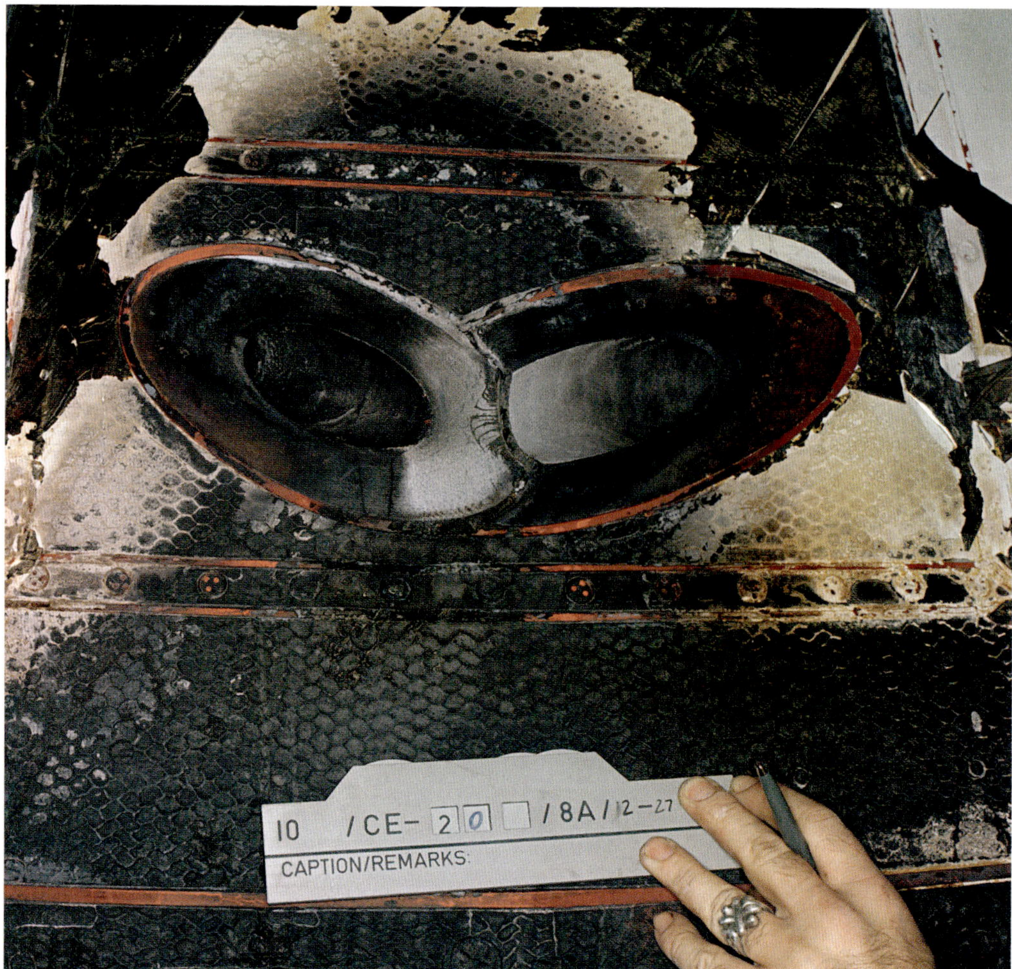

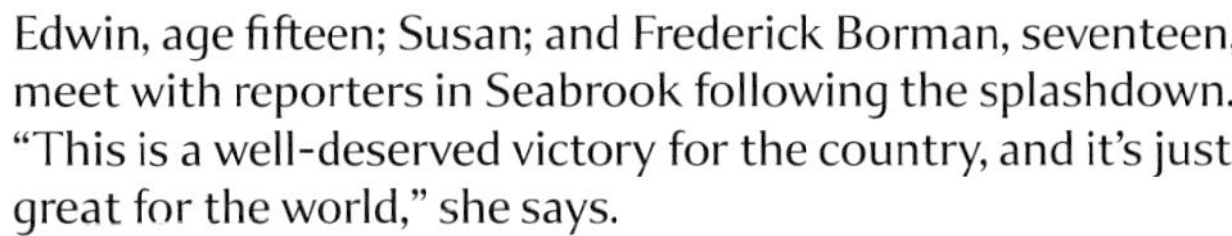

Edwin, age fifteen; Susan; and Frederick Borman, seventeen, meet with reporters in Seabrook following the splashdown. “This is a well-deserved victory for the country, and it’s just great for the world,” she says.

Jay, thirteen; Barbara, fifteen; Jeffrey, three; their mother, Marilyn; and Susan Lovell, ten, meet with reporters in Seabrook after splashdown. “I’m so proud for our country that our husbands could accomplish the mission,” she says. “There is not one word in the dictionary to describe how I feel today.”

Gayle, eight; Eric, four; Gregory, six; and Glen, ten, look on as their mother, Valerie Anders, speaks with reporters outside their home in Seabrook. “Giving thanks is a course from which we never graduate,” she says.

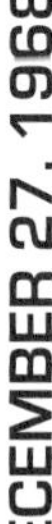

Ben James leads the crewmen from sick bay on the way to their quarters as the Marine honor guards salute.

The astronauts pause in Hangar Bay 3 to inspect the heat shield with James (*at left*) and Stonesifer (*at right*). The uprighting bags are still partially inflated.

The crewmen pose for photographers.

Anders, Lovell, and Borman enjoy a formal dinner that evening in the wardroom with *Yorktown*'s senior officers.

"To tell you the truth, I was just hoping we'd hit any ocean," jokes Anders.

"I've been living with two Air Force men for the past week, and it's great to be aboard," says Lovell.

"That reentry was really something," Borman says. "Boy, my eyes opened wide on that one."

Lovell uses a ceremonial sword to cut the traditional recovery cake on the hangar deck. He'd earlier give the oath of reenlistment to seven sailors.

The 21-foot-square, fourteen-layer cake is half chocolate and half white, is composed of sixty small sheet cakes, and weighs 540 pounds.

CHAPTER 13

December 28, 1968–May 7, 1969

The communications relay ship USS *Arlington* (AGMR-2) passes USS *Yorktown* near Hawaii on the morning of December 28. The modified aircraft carrier had been sent from the Gulf of Tonkin to serve as a floating switchboard and communication center for the recovery.

The astronauts greet *Arlington*'s sailors. The ship had been stationed about 9 miles away at splashdown.

NASA's Ben James (*left*) and Anders prepare to board a Grumman S-2 Tracker aircraft with Borman and Lovell for the flight from *Yorktown* to Hawaii. Anders had just been promoted to lieutenant colonel because of the mission but hasn't been informed. Borman and Lovell received their one-time promotions after Gemini VII.

At least seven thousand spectators and news media watch as the astronauts arrive at Hickam AFB near Pearl Harbor on Oahu at 2:30 p.m. HST. ABC uses the facilities of its Honolulu affiliate, KHVH-TV, to handle the network's TV pool responsibilities. The station has RCA TK-42 cameras.

Hawaii governor John Burns (D) (*left*) and MSC director of flight crew operations, Deke Slayton, lead the astronauts to the mics after they receive flowered leis and shake hands with Navy brass.

"I'm the only one of the Navy contingent among our crew," Lovell says, "and I had to tell them all what a wonderful place Hawaii was to be on a tour of duty here. Talking from the Navy point of view, I'm also glad to be part of the crew. We enjoyed our trip; we hope we contributed a lot toward our space program. It's a good program, but again, it's the people that are behind it that really make it count. And we appreciate our trip, and we appreciate your help and cooperation in making it possible."

The astronauts, carrying plaques from the governor, and Slayton pass three hula dancers from the Honolulu Hilton as they depart after about twenty minutes.

Susan Borman talks with a reporter on a cold, breezy early morning at Ellington AFB, northwest of MSC, before the astronauts arrive on December 29. Borman's father, Edwin, is on the left.

Wearing her new fur stole, Marilyn Lovell answers a reporter's questions with daughter Barbara behind her.

Airmen salute as the crewmen arrive at about 2:00 a.m. CST after a seven-hour flight aboard a Lockheed C-141 Starlifter aircraft.

MSC director Robert Gilruth greets Borman, with his son Edwin behind him. Anders and his wife, Valerie, are at the right.

The Anders family is reunited; Valerie wears a lei from her husband.

Borman presents his wife, Susan, with a lei.

A crowd of more than two thousand surrounds Gilruth and the crewmen on the runway as they approach the mics.

The astronauts wait as Gilruth introduces them. Astronaut Ed Mitchell (*hidden by a mic*) and his wife, Louise, are between Anders and Lovell.

"We want to thank you all for coming out so early in the morning or late at night," says Borman. "I guess there's no way to express, really, how great it is to be home, be with our family and friends again. It was a great flight, but we're really happy to be back here in Houston." Anders holds daughter Gayle.

"Let me also express my appreciation for all the wonderful turnout," Lovell says. "We had a wonderful trip, and we hope that you look toward the program like we do, because we think it's a great thing and it's going to go in great places shortly. We really had a great time." Valerie and Glen Anders are at left.

"We had a wonderful flight," adds Anders. "We're proud, as I hope you are, and I want to thank you as Americans for helping make it possible. Thanks a lot."

Borman

Anders signs an autograph after well-wishers break free from rope lines and swarm the astronauts, who comply with many requests for signatures. Valerie Anders is briefly separated by the surging crowd. The box on the next seat contains flowers from Hawaii.

Lovell, daughter Barbara, and Marilyn try to leave. "At two in the morning, I expected to get in my old blue bomb and go home," he quips. *AP photo*

Susan, Jim, Barbara, Marilyn, and Jay Lovell pose at their Seabrook home. *AP photo*

Anders admires a flag display made with Christmas lights at his home in Seabrook with children Alan (*left*) and Glen and Gayle. *AP photo*

The next afternoon, the crewmen enjoy a light moment during a tape-recorded debriefing at MSC. The sessions are scheduled for eight hours a day for two weeks.

The CM is offloaded from *Yorktown* at Ford Island in Pearl Harbor on December 29 for residual chemical remediation. Missing RCS panels were removed during safing.

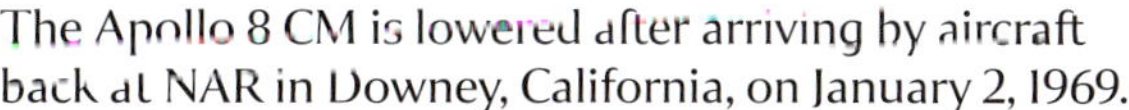

The Apollo 8 CM is lowered after arriving by aircraft back at NAR in Downey, California, on January 2, 1969.

NAR technicians prepare to seat the CM on a dolly. It will undergo two months of tests at NAR. *AP photo*

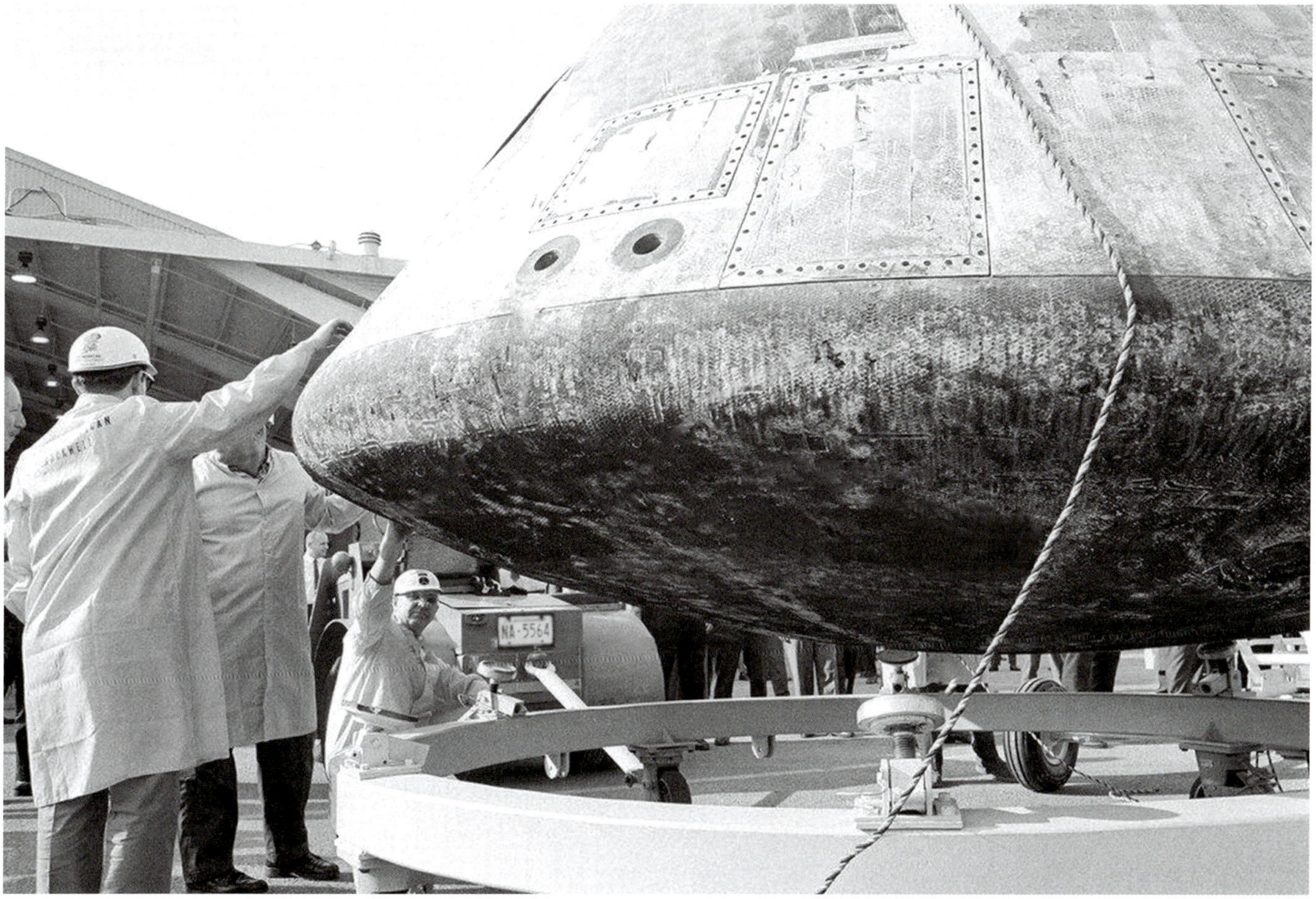

Marilyn Lovell is greeted by former NASA administrator Jim Webb at the White House before each of the astronauts receives the NASA Distinguished Service Medal on the morning of January 9, 1969. President Johnson's daughter Luci Johnson Nugent and grandson Lyn are at left, with Secretary of State Dean Rusk behind them at far left. It will be a busy day for the crew. *Bill Taub archive*

Left to right: Luci and Lyn, Marilyn Lovell, Jim Lovell, Lynda Johnson, Susan Borman, Frank Borman, and Lady Bird Johnson listen to Johnson (the Anders family is out of frame to the left). "These men represented in the vastness of space all mankind, all of its races, all of its nationalities, all of its religions, all of its ideologies," he says. "For seven days the earth and all who inhabit it knew a measure of unity through these brave men." *Bill Taub archive*

Anders receives his NASA Distinguished Service Medal, the highest award the agency can bestow.

Lovell accepts his medal. The ceremony is carried live by the three national TV networks.

Borman receives his medal.

Two-year-old Lyn Nugent breaks free from his mother and heads for his grandfather; Luci will scoop him up.

"We're grateful to this wonderful country," Borman says. "They've supported us in every way. And although we're symbolic of the country's greatness, we certainly feel very inadequate, and we're just very, very grateful."

Anders gives Johnson miniature copies of two recent treaties covering astronaut rescues and banning nuclear weapons in space that the crew carried on Apollo 8. *LBJ Library*

Lovell presents the president with an inscribed and signed Earthrise photo, which Borman describes as "a picture of the [LBJ] ranch," to laughter.

The crewmen, joined by Vice President Hubert Humphrey, lead a motorcade along Pennsylvania Avenue from the White House to the US Capitol.

In the House of Representatives chamber, Speaker John McCormick (D-Mass.) (*right*) leads applause after he asks the astronauts' families to stand. The crewmen will speak to a joint meeting of Congress about 12:30 p.m. EST. Humphrey is at left.

Valerie Anders; Marilyn Lovell; Susan Borman; Borman's parents, Edwin and Marjorie Borman; and his son Edwin Borman acknowledge the applause from the visitors' gallery. Borman's other son, Frederick, stands behind his brother.

Borman introduces Lovell by saying, "I was a little reluctant to do this because on the flight, every time I gave up the microphone, I seldom got it back." Lovell begins, "Distinguished people of the good earth . . ." He and Anders speak briefly. *Bill Taub archive*

"That one overwhelming emotion that we carried with us is the fact that we really do all exist on one small globe, and when you get out 240,000 miles, it really is not a very large earth," says Borman. The event is carried live by the national TV networks.

"Exploration really is the essence of the human spirit, and to pause, to falter, to turn our back on the quest for knowledge is to perish—and I hope that we never forget that," Borman concludes.

The Apollo 8 crew next holds a news conference at the State Department auditorium, also carried on live TV. They report they found proposed lunar-landing sites "sufficient and adequate," with lighting "much more amenable than we had thought."

Borman announces that Apollo 8 would be his last space flight. He'll become the number 2 manager of the Astronaut Office at MSC. The binder has photos for a slide show that Anders will narrate.

Later that day, NASA announces the Apollo 11 crew and says that Lovell will be the backup commander, with Anders as backup CM pilot.

The astronauts wrap up their day with a private reception and dinner at the National Air and Space Museum. Charles Lindbergh's Spirit of St. Louis is suspended above. The edge of Friendship 7, John Glenn's Mercury spacecraft, is visible at left.

The following day, January 10, New York City mayor John Lindsey presents each crewman with the city's Medal of Honor on the City Hall steps at 12:30 p.m. after a ticker tape parade up Broadway in subfreezing temperatures. "They have achieved the impossible, conquered the unknown, and shown the world what this country is made of and what it can do," he says. New York governor Nelson Rockefeller is to the left of Lovell. Next is a luncheon at the New York State Theater at Lincoln Center, and they later receive an ovation at the United Nations. *Bill Taub archive*

The day ends with a roast beef dinner in the Waldorf Astoria's Grand Ballroom attended by 2,500 political leaders and guests. *Bill Taub archive*

Valerie Anders; Anders; Secretary of State William Rogers; Susan Borman; Borman; Rockefeller; his wife, Happy Rockefeller; and Lovell are at the head table. *Bill Taub archive*

The astronauts are besieged with autograph requests for almost an hour.

Lovell signs dinner programs. He says he'll leave the dinner "slightly incapacitated. With my left hand, I will not be able to write. With my right hand, I won't be able to shake hands." *Bill Taub archive*

Anders signs autographs as Rockefeller looks on (*at left*). "Ladies and gentlemen, this started out as a state dinner and is turning into a love-in," the governor shouts. "Will you please take your seats!" *Bill Taub archive*

The next morning, before leaving for Miami, the astronauts tape an episode of the CBS News program *Face the Nation* for broadcast the following day. Asked about the mission's cost, Borman replies, "It doesn't trouble me at all. I think that what we've done in the Apollo program and in the space program is absolutely essential. I look at it as technical insurance for the future of this country, and I don't think we can afford to neglect it."

The astronauts lead the Pledge of Allegiance on the field before Super Bowl III on January 12 at the Orange Bowl in Miami, Florida. *Bill Taub archive*

Borman's son Edwin serves as a ball boy, as does his brother. Both had done the job for the Houston Oilers. *Bill Taub archive*

The astronauts and their families have 50-yard-line seats. "I'm kind of partial to the American League," says Borman, "but I'm also a great fan of [Baltimore quarterback] Johnny Unitas." The AFL's New York Jets upset the NFL's Baltimore Colts 16–7.

Borman, Lovell, and Anders with Hugh Downs, host of NBC's *Today* show

The next day, January 13, the Anders family rides through downtown Houston as an estimated quarter of a million people line thirty-six blocks for a "welcome home" parade. *70 mm still frame*

The Lovells acknowledge the cheers amid confetti and streamers. *Bill Taub archive*

Borman waves to well-wishers. NASA photographer Bill Taub (*back to camera*) is at far left.

Lovell presents Houston mayor Louie Welch with an Earthrise photo he calls "a picture of Houston." Welch holds a flown Texas flag and Apollo 8 patch he received from the crew. Texas governor John Connally, who receives the same framed presentation, is behind Welch on the reviewing stand at the Albert Thomas Convention Center. The astronauts wear gold valor medals just presented by Welch; they are the first to receive the Houston award.
Bill Taub archive

The crewmen (*out of frame*) leave empty seats as they stand for a marching unit. *Left to right*: Kurt Debus, George Mueller, Valerie Anders, Marilyn Lovell, Susan Borman, Sen. Ralph Yarborough (D-Texas), and Gilruth. They later attend a ceremony at MSC where ninety-seven government, military, and contractor personnel receive NASA medals.

The next day, January 14, an estimated one and a half million people turn out for a noontime parade through the central business district in Chicago, Illinois.

The astronauts are showered with confetti and ticker tape. “When people turn out in cold like this,” says Anders, “you know who your friends are.” The temperature is 25 degrees Fahrenheit.
Bill Taub archive

The crowd cheers the astronauts. "Thank you!" many shout. "God bless you!" *Bill Taub archive*

Mayor Richard Daley (*right*) makes the astronauts honorary citizens of Chicago during a special City Council session. "The accomplishments of these three young men underscore something that this nation has needed," he says. "Confidence in ourselves." *Bill Taub archive*

The astronauts are next honored at a luncheon in the Conrad Hilton's International Room.

Anders holds up a framed US flag that was carried to the Moon, which he presents to Daley. The crewmen then field questions from 3,500 high school students.

On January 20, the Bormans and Lovells watch President Richard Nixon's inaugural parade from the viewing stand at the White House. Former Rhode Island governor John Chaffee (*behind Susan Borman*) will soon be nominated to be Navy secretary by Nixon.

The astronauts meet with Nixon at the White House on January 30, before Borman and his family leave on a goodwill tour of eight European countries. The trip, the president says, will emphasize that "the knowledge which made possible these great discoveries is not limited to this nation. No nation has a monopoly on that kind of knowledge." Lovell and Anders will not go, since both of them are on the Apollo 11 backup crew, but the three crewmen and their wives stay for dinner and later watch movies from the flight.

With the Collier Trophy behind them, the astronauts hold a small replica on May 7, 1969, in Washington, DC. The crewmen receive the annual award for the greatest US achievement in aeronautics or astronautics in 1968. NASA has just assigned Borman to head a space station task group.

Epilogue

Frank Borman, who has just returned from the Soviet Union, speaks with reporters at the LC-39 Press Site on July 12, 1969. Asked what the goal was of an unmanned Russian lunar probe just launched, he speculates it is a sample return mission. “It would be a great feat if they can beat us unmanned,” he says. “I think it is unlikely it would take the edge off Apollo 11.” KSC public affairs chief Jack King is at right.

Borman, NASA's White House liaison, and President Nixon attend a Washington Senators baseball game at Robert F. Kennedy Memorial Stadium in Washington, DC, on July 15, 1969, the night before the Apollo 11 launch. Nixon had originally planned to be at KSC for a dinner with the astronauts, but NASA physicians canceled it because the crew was in medical quarantine, a decision that Borman criticizes. *AP photo*

Borman and Nixon greet the Apollo 11 astronauts aboard USS *Hornet* on July 24, 1969. "Frank Borman says you're a little younger by reason of having gone into space," Nixon says. "Is that right?" CM pilot Mike Collins replies, "We're a lot younger than Frank Borman."

Apollo 11 backup LM pilot Fred Haise (*left*) and backup commander Jim Lovell collect specimens during geology training near Sierra Blanca and the ruins of Fort Quitman, about 90 miles southeast of El Paso, in West Texas, on February 25, 1969. Haise uses a Sony TC-110 cassette recorder to take notes.

Borman announces his retirement from NASA and the USAF on January 29, 1970, to become vice president of Electronic Data Systems, a computer services company, and to start a civic foundation with its owner, Dallas millionaire H. Ross Perot. He remains a NASA consultant on space stations.

Borman joins Eastern Airlines on July 1, 1970, and becomes chief executive officer in 1975. He resigns in June 1986 after layoffs and pay cuts fail to make the airline profitable.

Left to right: Former astronauts Alan Shepard, Gus Grissom (represented by his widow, Betty), John Glenn, Pete Conrad, Borman, and Neil Armstrong each receive the first Congressional Space Medal of Honor from President Jimmy Carter (*far right*) at a ceremony in the VAB on October 1, 1978. The vertical stabilizer of space shuttle test article *Pathfinder* is behind them.

Lovell waits to enter Apollo 11's LM-5 during an altitude chamber test in the MSOB on March 19, 1969.

Lovell (*left*) and Anders (*right*) flank Vice President Spiro Agnew during the Apollo 12 countdown in the Management Operations Room in Firing Room 2 at LC-39 on November 14, 1969. Al Siepert, KSC deputy director, is next to Lovell.

Lovell prepares for EVA training as Apollo 13 commander in the Flight Crew Training Building at KSC on January 28, 1970.

Apollo 13 astronauts Ken Mattingly, Lovell, and Haise at Pad 39A on April 6, 1970. Mattingly, suspected of measles exposure, would be replaced by Jack Swigert as LM pilot three days later.

Lovell with President Nixon during a brief welcoming ceremony at Honolulu International Airport in Hawaii on April 18, 1970, after Apollo 13's lunar-landing mission was aborted. The astronaut wears his Presidential Medal of Freedom, just awarded by Nixon.

Anders signs autographs at the VIP stands near the VAB before the Apollo 9 launch on March 3, 1969.

Anders, as Apollo 11 backup CM pilot, trains in the centrifuge in Building 29 at MSC on April 15, 1969.

Anders waits in the VIP stands for the launch of Apollo 11 on July 16, 1969, with astronaut Tom Stafford (*left*), and Siepert (*right*). Behind them are Lady Bird Johnson, former president Johnson, Agnew, and his wife, Judy.

Anders and Agnew follow the launch of Apollo 12 in Firing Room 2 in the LCC on November 14, 1969. Lovell is at left.

Anders (*left*) and astronaut William Pogue attend the launch of the Skylab workshop on May 14, 1973, at the VIP area.

Anders in 1974 as the first chairman of the Nuclear Regulatory Commission. He would later work for GE and Textron before becoming chairman of General Dynamics in 1991.

On September 29, 1971, a crane raises the Apollo 8 CM and its dolly at the Museum of Science and Industry in Chicago.

The CM is lowered through an opening cut in the ceiling.

The CM reaches the floor.

The CM, on loan from the Smithsonian Institution, on display in 1994.
Photo by J. L. Pickering

Front view of the CM.
Photo by J. L. Pickering

Borman's pressure suit on display in 1994 at the Chicago museum.
Photo by J. L. Pickering

On October 6, 2018, the astronauts appear at the thirty-eighth Columbian Ball at the museum, an annual fundraiser. During the evening, the International Astronomical Union officially names three lunar features in their honor: "Mount Marilyn," for Lovell's wife, and craters "Anders' Earthrise" and "8 Homeward" for Borman. *Photo by Mark Usciak*

Borman responds during a question-and-answer session with journalist Bill Kurtis. Borman died in 2023. *Photo by Mark Usciak*

Anders, who died in 2024. *Photo by Mark Usciak*

Lovell, who died in 2025. *Photo by Mark Usciak*

Abbreviations

AFB	air force base
AFS	air force station
AMS	Apollo Mission Simulator
AS	Apollo-Saturn
BP	boilerplate
CDDT	Countdown Demonstration Test
CM	command module
CSM	command/service module
CST	Central Standard Time
EST	Eastern Standard Time
EVA	extravehicular activity
HST	Hawaii Standard Time
ILC	International Latex Corp.
IU	instrument unit
KSC	Kennedy Space Center
LC	launch complex
LCC	Launch Control Center
LES	launch escape system
LM	lunar module
LOI	lunar orbit insertion
LOX	liquid oxygen
LTA	lunar module test article
LUT	launch umbilical tower
ML	mobile launcher
MCC	Mission Control Center
MOCR	Mission Operations Control Room
MSC	Manned Spacecraft Center
MSFC	Marshall Space Flight Center
MSOB	Manned Spacecraft Operations Building
MV	motor vessel
NAA	North American Aviation
NAR	North American Rockwell
NASA	National Aeronautics and Space Administration
RCS	reaction control system
SA	Saturn-Apollo
SLA	Spacecraft/Launch Vehicle Adapter
SM	service module
SPS	service propulsion system
TLI	translunar injection
TEI	trans-Earth injection
US	United States
USAF	United States Air Force
USN	United States Navy
VAB	Vehicle Assembly Building

Bibliography

"Analysis of Apollo 8 Photography and Visual Observations." NASA SP-201. Compiled by NASA Manned Spacecraft Center, Office of Technology Utilization, Washington, DC, 1969.

Apollo 8 Mission Evaluation Team. "Apollo 8 Mission Report." MSC-PA-R-69-1. Houston, TX: Manned Spacecraft Center, February 1969.

"Apollo 8 Recovery/Cruise Around the Horn." USS *Yorktown* CVS-10. October 1968–February 1969.

Apollo Program Office–MAO. Apollo 8 Mission. "Post Launch Mission Operation Report." Report M-932-68-08. Washington, DC, 1969.

Benson, Charles D. *Moonport: A History of Apollo Launch Facilities and Operations*. Washington, DC: Scientific and Technical Information Office, National Aeronautics and Space Administration, 1978.

"Big Receiver Picks Up Tiny Signal." *Broadcasting* 75, no. 27 (December 30, 1968). Washington, DC: Broadcasting Publications.

Brooks, Courtney G., and Ivan D. Ertel. *The Apollo Spacecraft: A Chronology*. Vol. 3. Washington, DC: National Aeronautics and Space Administration, 1976.

Chaikin, Andrew. "Who Took the Legendary Earthrise Photo from Apollo 8?" *Smithsonian Magazine*, January 2018.

Coan, Paul P. "Apollo Experience Report: Television System." NASA Technical Note D-7476. Houston, TX: National Aeronautics and Space Administration, Lyndon B. Johnson Space Center, 1973.

DOD Manned Spacecraft Recovery Forces and NASA-MSC Landing and Recovery Division. "Apollo Recovery Operational Procedure Manual." MSC-01856. Houston, TX, 1968.

Gibson, Walter G. "A Scan Converter for Apollo Television." *RCA Engineer* 11, no. 6 (April–May 1966). Princeton, NJ: Radio Corporation of America.

Godwin, Robert, and Alan Lawrie. *Saturn V: The Complete Manufacturing and Test Records*. Burlington, ON: Apogee Books, 2005.

Kozloski, Lillian D. *US Space Gear: Outfitting the Astronaut*. Washington, DC: Smithsonian Institution Press, 1994.

Morse, Mary Louise, and Jean Kernahan Bays. *The Apollo Spacecraft: A Chronology*. Vol. 2. Washington, DC: National Aeronautics and Space Administration, 1973.

National Aeronautics and Space Administration. "Apollo 8 PAO Mission Commentary." Houston, TX: Manned Spacecraft Center, 1968.

National Aeronautics and Space Administration. "Apollo 8 Press Kit." Release 68-208. Washington, DC, December 1968.

National Aeronautics and Space Administration, Launch Operations Center. "NASA Space Utilization at AMR." Cape Canaveral, FL, 1962.

National Aeronautics and Space Administration. *Apollo Recovery Operational Procedures Manual, Revision C*. Houston, TX: Manned Spacecraft Center, Landing and Recovery Division, 1971.

North American Rockwell. *Apollo Spacecraft Familiarization Manual*. Downey, CA, 1967.

North American Rockwell, Space Division. *Apollo Spacecraft News Reference*. Downey, CA, 1969.

Pavlosky, James E., and Leskie G. St. Leger. "Apollo Experience Report Thermal Protection Subsystem." Houston, TX: National Aeronautics and Space Administration, Lyndon B. Johnson Space Center, 1974.

Phinney, William C. "Science Training History of the Apollo Astronauts." NASA SP-2015-626.

Saturn V Flight Evaluation Working Group. "Saturn V Launch Vehicle Flight Evaluation Report–AS-503 Apollo 8 Mission." Huntsville, AL: National Aeronautics and Space Administration, George C. Marshall Space Flight Center, 1969.

Slovinac, Patricia. "Cape Canaveral Air Force Station, Launch Complex 39, HAER No. FL-8-11-E Altitude Chambers." Atlanta, GA: Historic American Engineering Record, National Park Service Southeast Region, Department of the Interior, 2009.

Ward, Jonathan L. *Rocket Ranch: The Nuts and Bolts of the Apollo Moon Program at Kennedy Space Center*. Chichester, UK: Springer Praxis Books, 2015.

Index

BORMAN LOVELL ANDERS